NEWTON ON MATHEMATICS AND SPIRITUAL PURITY

ARCHIVES INTERNATIONALES D'HISTOIRE DES IDÉES

INTERNATIONAL ARCHIVES OF THE HISTORY OF IDEAS

183

NEWTON ON MATHEMATICS AND SPIRITUAL PURITY

by

AYVAL LESHEM

NEWTON ON MATHEMATICS AND SPIRITUAL PURITY

by

AYVAL LESHEM

Hebrew University, Jerusalem

KLUWER ACADEMIC PUBLISHERS

DORDRECHT / BOSTON / LONDON

A C.I.P. Catalogue record for this book is available from the Library of Congress.

ISBN 1-4020-1151-2

Published by Kluwer Academic Publishers,
P.O. Box 17, 3300 AA Dordrecht, The Netherlands.

Sold and distributed in North, Central and South America
by Kluwer Academic Publishers,
101 Philip Drive, Norwell, MA 02061, U.S.A.

In all other countries, sold and distributed
by Kluwer Academic Publishers,
P.O. Box 322, 3300 AH Dordrecht, The Netherlands.

Printed on acid-free paper

Printed in the Netherlands.

To my father,
Shlomo Leshem, and
to the memory of my
late mother, Nurit
Leshem.

TABLE OF CONTENTS

PREFACE

I have been writing and revising this book for the past seven years, changing and transforming it many times, always for the same reason. After I had been able to express the abstract idea I had in mind I would reflect about it and discover that something deeper lay beyond. If professional demands had not obliged me to produce a final manuscript I would never have finished the writing. If I had been asked to enumerate all the obstacles I faced whilst turning this book into print another and perhaps no less intriguing book would have resulted. I am aware of my good fortune in having finally finished the task, and I would like to thank the many people whose help was invaluable.

First and foremost I would like to dedicate this book to my late merciful and most beloved mother, Nurit Leshem, and to my wonderful and wise father, Shlomo Leshem, who learned to accept me as I am. My mother's unexpected death and the many existential questions that it presented to me profoundly modified the way I used to think about intellectual matters. It brought me to the strong realization that beyond the scientific and intellectual realm a very simple spiritual truth exists that shapes the way we think. This truth was known to scholars of the seventeenth century, such as Newton and Leibniz, but today in the academic and professional world, to which I belong, it is considered by most as a categorical mistake and a non-sophisticated discourse. Even before my mother passed away I was exposed to this realm when exploring Leibniz's science (in my Ph.D. thesis), yet my reflections remained in the regular intellectual channels. In those years the late Prof. Amos Funkenstein planted the seeds in my mind for the many fruits that have ensued. Anyone who was as fortunate as I was to study with Amos can see throughout the pages of this book how much influence he had upon the way I think today.

Many thanks to Mary Terrall, Tom Ryckman, Ted Porter, Ken Alder, Tzion Asoulin, Arie Issar and Amalya Oliver, who believed in the value of my work when I was not sure of its validity. Thanks also to Steve Snobelen and Allison Coudert, who gave me selflessly their time and wisdom all along. I benefited also from the help of Michael Ben-Chaim, Matt Goldish, Avihu Zakai, Michael Heyd, Benny Shanon, Yemima Ben-Menahem, and Scott Mandelbrote. Special and warm thanks to Yakir Shoshani, Aviah Luz and Limor Hacklay, who listened attentively and made illuminating suggestions concerning ideas that kept on appearing as the work reached its final form.

I would also like to thank my many students whose affection and appreciation showed me that the content and the way I present the ideas appearing in this book are important to their generation. Invaluable were also the warm and wise hearted suggestions of my sister, Sharon Leshem-Zinger, and the far-sighted insights of my brother, Dotan Leshem. I value the computer expertise that Guy Gavron and Hanoch Segal offered me. Their virtuosity with the computer and the stimulating discussions we had around it also inspired certain spiritual and intellectual realizations as presented in this book.

I am grateful to Mira Frankel Reich for her patient and meticulous editing, without which the book would not have become as communicative as it did, and to Raviv Shalev for the help in preparing the manuscript for print. Needless to say, none of the above mentioned bears any responsibility for the mistakes, misprints and misconceptions, which may remain. I owe warm thanks to the anonymous readers of the text, to Kluwer Academic Publishers for supporting the publication, and to a fellowship from Yad Hanadiv for funding the editing of the manuscript. I also gratefully acknowledge the Jewish National and University Library, Jerusalem and the Provost and Fellows of King's College, Cambridge for permission to quote from manuscripts in their archives.

Finally I am indebted to Newton and Leibniz and above all to God who made me the gift of this intellectual and spiritual journey. I am also deeply appreciative of the support of Offer Ramati (and his family) during my research and of the love, clarity and creative imagination of my daughter, Adi; without her I might have forgotten why the search for truth is so important.

CHAPTER I

THE SEARCH FOR TRUTH

"[Truth is] the offspring of silence and unbroken meditation." (Newton, Keynes MS. 130, p. 7.)

"It is in Newton's belief in the unity of Truth," wrote the late Betty Jo Tetter Dobbs, "guaranteed by the unity and majesty of God, that one may find a way to reunite his many brilliant facets, which however well polished, now remain incomplete fragments."[1] Newton believed that the truth can be uncovered only by "a remnant, a few scattered persons which God has chosen, such as without being led by interest, education, or humane authorities can set themselves sincerely & earnestly to search after [it]"[2] Such a "Truth," writes Newton, "is ever to be found in simplicity, and not in the multiplicity and confusion of things."[3] But what does this simple truth consist of for Newton?

---

[1] Betty Jo Teeter Dobbs, *The Janus Faces of Genius: The Role of Alchemy in Newton's Thought*, (Cambridge, 1991), p. 18. See also, p. 247; idem, "Newton as Alchemist and Theologian," N.J.W. Thrower (ed.), *Standing on the Shoulders of Giants*, (Berkeley, 1990), pp. 129-140. On the unity of Newton's thought, see also: J.E. McGuire and P.M. Rattansi, "Newton and the Pipes of Pan," *Notes and Records of the Royal Society of London*, vol. 21 (1966), pp. 108-143.

[2] Newton, "Treatise on the Apocalypse," Jewish National and University Library, Jerusalem, Yahuda MS. 1, fol. 1, cited in Frank E. Manuel, *The Religion of Isaac Newton*, (Oxford, 1974), p. 108, and also in James E. Force, "Newton's 'Sleeping Argument' and the Newtonian Synthesis of Science and Religion," *Standing on the Shoulders of Giants*, pp. 118-19.

On Newton's view on the chosen remnant, see also: Matania Z. Kochavi, "One Prophet Interprets Another: Sir Isaac Newton and Daniel," J. Force and R. Popkin (eds.), *The Books of Nature and Scripture Recent Essays on Natural Philosophy, Theology, and Biblical Criticism in the Netherlands of Spinoza's Time and the British Isles of Newton's Time*, (Dordrecht, 1994), pp. 105-122; Scott Mandelbrote, "Isaac Newton and Thomas Burnet: Biblical Criticism and the Crisis of Late Seventeenth-Century England," *The Books of Nature and Scripture*, pp. 149-178, 158; idem, "'A Duty of the Greatest Moment': Isaac Newton and the Writing of Biblical Criticism," *The British Journal for History of Science*, vol 26 (1993), pp. 281-302, 289; Stephen D. Snobelen, "Isaac Newton, Heretic: the Strategies of a Nicodemite," *The British Journal for History of Science*, vol. 32 (1999), pp. 381-419; idem, "'God of gods, and Lord of lords': the Theology of Isaac Newton's General Scholium to the *Principia*," *Osiris*, vol. 16 (2001), pp. 169-208; Robert Markley, "Newton, Corruption, and the Tradition of Universal History," J.E. Force and R. Popkin (eds.), *Newton and Religion: Context, Nature and Influence*, (Dordrecht, 1999), pp. 121-145, 122, 128-9; Richard S. Westfall, *Never at Rest*, (Cambridge, 1998), p. 325.

[3] Newton, *Rules for Methodizing the Apocalypse*, Yahuda MS. 1.1, reprinted in Manuel, *The Religion of Isaac Newton*, p.120. On the deep implications of the notion of simplicity for his natural philosophy and interpretations of prophecy, see: James E. Force, "Newton's God of Dominion: the

This book will argue that Newton's youthful method of fluxions, though dealing with abstract mathematics, had a definitive role in making clear to him his notion of the true simple designed order that God imprinted upon matter in creation and from then on has sustained from day to day. The mathematical method captures and exemplifies how God sustains the material designed order through his absolute space and time, which are a source of unlimited creative energy. It also elucidates the mechanism of the natural decline of all processes in nature. This simple method is thus analogous in structure to the simple creed, which God re-instilled throughout history in the religious founders and prophets who recall human beings to the true worship of God.

Scholars who have classified Newton's works in two separate categories, one scientific and one esoteric, miss the depth and richness of his pious search for the divine truth. "The theory of two Newtons," writes David Castillejo, "has to be discarded," because though Newton's "attention is multiple, [and] not linear, his system is absolutely unitary."[4] Newton pursued the truth hidden in nature (physics and alchemy), in Scripture, prophecy, and in the actions of providence in human history. In all of these fields a common zeal and theme appeared - a quest for the simple machinery through which God creates, governs, sustains, and replenishes creation. This search was not undertaken for the sake of intellectual pleasure alone, but was an endeavor to reveal God's hidden work in creation and his ways of sustaining and replenishing the simple created harmonic order.

"It is clear," writes Richard H. Popkin, "that Newton held that God was the creator, designer and sustainer of the world." Newton believed that the Bible was "the way God has communicated [truth] to us in words, just as Nature is God's communication in things." Yet both the "natural and verbal messages require tremendous effort, insight, and pious attention to understand [them]."[5] Or, as Richard S. Westfall has put it, "the correspondence of prophecy with fact demonstrated [for Newton] the dominion of God, a dominion exercised over human history even as it [was] exercised over the natural world."[6] The Bible was "not a revelation of mysteries beyond human reason,... but an historical account of God's dominion, [which was] meant to demonstrate his omnipotent power to men, as Nature demonstrates his infinite

---

Unity of Newton's Theological, Scientific, and Political Thought," J. Force and Richard H. Popkin (eds.), *Essays on the Context, Nature, and Influence of Isaac Newton's Theology* (Dordrecht, 1990), pp. 75-102; and Snobelen, "God of gods."

[4] David Castillejo, *The Expanding Force in Newton's Cosmos*, (Madrid, 1981), pp. 15, 78.

[5] Richard H. Popkin, "Newton's Biblical Theology and his Theological Physics," P.B. Scheurer & G. Debrock (eds.), *Newton's Scientific and Philosophical Legacy*, (Dordrecht, 1988), p. 91.

[6] Westfall, *Never at Rest*, p. 329; on the connection between Newton's theology and natural philosophy see also: Richard S. Westfall, "Newton's Theological Manuscripts," Z. Bechler (ed.), *Contemporary Newtonian Research*, (Reidel, 1982), pp. 129-143.

wisdom."[7]

James E. Force adds to this dual aspect of truth, expressed in Nature and in Scripture, that "Newton's thought as a whole presents a systematic unity of startling coherence, [and] the key to understanding its uniquely integrated quality, is his conception of the nature of the Lord God of Dominion." Indeed, Newton's "extremely voluntaristic notion of the Lord God of Israel, a God of total power and absolute 'dominion' affects every aspect of his metaphysics including his approach to a scientific understanding of Nature."[8] All of Newton's studies contribute to the same concept regarding the dominion of the "Lord and Master" over all aspects of creation, because "truly for Newton, natural philosophy and religion are indeed synthesized in a most 'holy alliance.'"[9] The idea which lies behind all secrets of creation, writes Newton, is that "God, the wisest of beings require[s] of us to be celebrated not so much for his essence as for his actions the *creating, preserving, and governing* of all things according to his good will and pleasure."[10] This God of Dominion creates the world and then continuously preserves creation through secondary mechanical causes or through "extraordinary, direct voluntary interposition of his will." God first created matter, and then "install[ed] the laws which regulate it, continually supervis[ing] the maintenance and repair of those laws, and occasionally suspend[ing] them."[11] As time passes the state of the world continues to deteriorate due to material interactions and the indulgence of human beings in idolatry. God then periodically repairs and restores the original pure and divine state of creation with the assistance of material agents, like comets,[12] and messenger souls (as will be discussed below). This process of the deterioration of the order of the world is better exemplified in world history than in nature though the same process occurs in both spheres.

---

[7] Westfall, "Newton's Theological Manuscripts," p. 138.

[8] James E. Force, "Newton, the Lord God of Israel and Knowledge of Nature," Richard H. Popkin and G.M. Weiner (eds.), *Jewish Christians and Christian Jews: From the Renaissance to the Enlightenment* (Dordrecht, 1994), pp. 131-58, 133, 132. See also: idem, "Newton's God of Dominion", pp. 75-102; idem, "Samuel Clarke's Four Categories of Deism, Isaac Newton and the Bible," R. Popkin (ed.), *Skepticism in the History of Philosophy* (Dordrecht, 1996), pp. 53-74; idem, "The Nature of Newton's 'Holy Alliance' between Science and Religion: From the Scientific Revolution to Newton (and back again)," Margaret J. Osler (ed.), *Rethinking the Scientific Revolution* (Cambridge, 2000), pp. 247-70.

[9] Force, "Newton's God of Dominion", p. 151.

[10] Newton, Yahuda MS. 21, fol. 1 recto, cited in Force, "Newton's God of Dominion," p. 85.

[11] Force, "Newton's God of Dominion", pp. 87, 91.

[12] On this issue, see: Sara Schechner Genuth, "Newton and the Ongoing Teleological Role of Comets," *Standing on the Shoulders of Giants*, pp. 299-311; idem, *Comets, Popular Culture and the Birth of Modern Cosmology* (Princeton, 1997); Simon Schaffer, "Comets & Idols: Newton's Cosmology and Political Theology," P. Theerman and A. Seeff (eds.), *Action and Reaction: Proceedings of a Symposium to Commemorate the Tercentenary of Newton's **Principia*** (London, 1992), pp. 206-231.

One of Newton's "central theological concerns throughout his life [was] to combat what he calls idolatry." According to James Force, to Newton "worshipping anything but the Lord God of true and supreme dominion lessens the absolute nature of God's dominion and constitutes idolatry."[13] Newton sought all the examples he could find of idolatry both from the classical world and from the period before Christ.[14] Matt Goldish argues that in Newton's mind, "God revealed a true, simple religion to Adam, which consisted of two elements: love of God and love of man. These were not nebulous conceptions, but they rather involved specific commandments, such as not worshipping false gods, or the Devil, not lusting after material achievement, and not quarreling over metaphysical points which do not bear on this simple creed. Along with this Ur-religion, God also revealed certain wisdoms of natural philosophy, which were to remain independent of the two essential religious commandments."[15] This Ur-religion was a Vestal religion that has left traces in the record of all known civilizations. The pure worship of God was "performed in vestal temples (prytanaea) which were designed as microcosms centered around a holy fire."[16] According to Robert Iliffe, this Vestal religion made a close connection between theology and natural philosophy since "men were supposed to worship their God by the study of 'the frame of the world.' Thus all over the learned world, worship was organized by the local equivalent of the priest."[17] Noah and his sons propagated this worship but soon the original worship was corrupted in Egypt and elsewhere into an idolatrous hero and ancestor worship. Noah's descendants were first "idolized by their people as gods and ultimately identified with stars, planets, terrestrial elements, and mythological creatures recognized through various symbols."[18] This state of deterioration did not last for long. In merit of his recognition of the true God, Abraham was vouchsafed another opportunity to restore the right religion, "but it was constantly tainted by outside influences, until God sent Moses to teach

---

[13] Force, "Newton's God of Dominion", p. 80; see also, Markley, "Newton, Corruption, and the Tradition of Universal History," pp. 122-3.

[14] Rob Iliffe, "'Those 'Whose Business It Is To Cavill': Newton's Anti-Catholicism," *Newton and Religion*, pp. 97-121, 123.

[15] Matt Goldish, *Judaism in the Theology of Sir Isaac Newton*, (Dordrecht, 1998), pp. 11-12.

[16] Goldish, "Newton on Kabbalah," Force and Popkin (eds.), *The Books of Nature and Scripture*, pp. 89-103, 92. For a precise and short account of the history of the original religion, see: Westfall, "Newton's Theological Manuscripts." A longer account on this history appears in: Kenneth Knoespel, "Interpretive Strategies in Newton's *Theologiae Gentilis Origines Philosophiae*," *Newton and Religion*, pp. 193-4.

[17] Robert Iliffe, "'Is He Like Other Men?' The Meaning of the **Principia Mathematica**, and the Author as Idol," Gerald Maclean (ed.), *Culture and Society in the Stuart Restoration*, (Cambridge, 1995), pp. 165, 166, 167. On the Vestal religion, see also: Force, "Samuel Clarke's Four Categories of Deism," pp. 56-67; Force, "Newton, the "Ancients," and the "Moderns," *Newton and Religion*, pp. 237-59, 253-4.

[18] Force, "Samuel Clarke's Four Categories of Deism," p. 57.

[once again] the correct Noachian worship to the Jews." Following this, the Old Testament became a story of the corruption of the same simple religion through which the Jews "repeatedly fell to the influence of surrounding nations' contaminated worship, for which they were punished by exile." Jewish prophets kept restoring the truth since the Israelites were the chosen people. Nonetheless, in the long run the Israelites did not follow the messages given to them by the prophets so "Jesus came as a prophet to restore the correct religion, whose essential tenets, love of God and fellowman, are the same in Christianity and Judaism."[19] Following Christ's message, the history of Christianity is described by Newton as a similar pattern of corruption through idolatry, the re-instillation of some of the truth of the simple creed of Noachide worship, followed once again by corruption, and once again instillation and corruption. "It is only in the Last Times," writes Goldish, "the times when Newton lived, that God granted the tools of the scientific method to certain [wise] men, in order that they might reveal the workings of Providence in history which are encoded already in the prophets, mainly in Moses' warning and Daniel and the book of Revelation."[20]

Nonetheless, this simple truth encoded in prophecy and in nature was to be revealed and discovered only by a chosen remnant.[21] This kind of interpretation of prophecy "was conditional on a spiritual capability and, hence, [on] the nature of the faith of the exegete." "Newton developed a sense of chosenness," says Matania Z. Kochavi, "owing to his scientific understanding, and employed [this] knowledge in the process of interpreting" prophecy.[22] Even in his historical writings, writes Robert Markley, "Newton sketches a fragmentary history of the Fall into idolatry and offers at least glimpses of his program for reinvigorating a corrupt world: religious purity, techno-scientific knowledge, and dedication of a handful of individuals committed to demonstrating scientifically, theologically and historically the interlocking principles that can sustain an unsullied faith."[23]

We see then that Newton's multi-faceted work was directed towards what was to him a most simple truth regarding the work of providence in world history. From the beginning of creation, as time passes, a process of decay and corruption in humanity causes believers to deviate from the original pure Vestal religion (as well as all the following acts of restoration), to stop following the original divine path of simplicity, love and truth and begin to adopt contorted and distracted ways, which Newton identifies with idolatry. Thus as we move in world history away from acts of restoration of the original pure worship of the Lord and Master of creation, we find that human beings'

[19] Matt Goldish, "Newton on Kabbalah," pp. 89-103, p. 92.

[20] Matt Goldish, *Judaism*, pp. 11-12.

[21] On Newton and the remnant class, see note (2).

[22] Matania Z. Kochavi, "One Prophet Interprets Another," pp. 105-123, 105, 118.

[23] Markley, "Newton, Corruption, and the Tradition of Universal History," p. 123. On this issue, see also: Westfall, "Newton's Theological Manuscripts," p. 138.

understanding of God's simple truth becomes distorted, confused and complicated. Nevertheless, throughout human and natural history, in all of God's acts of restoration the same original truth is re-instilled. The new restoration in turn becomes corrupted, and a new re-instillation of truth is required. This process of re-instilling a purer and more spiritual order for the world, followed by a natural process of decay, can be observed in the world history of humanity as well as in the history of the material world. Without God's dominion, deterioration and corruption would have destroyed humanity and the material world long ago. Indeed, God needs to replenish the solar system as well, in this case with comets, otherwise the solar system would collapse.

Newton's contribution to science is well known; his deep interest and research in ancient Jewish science is less familiar. Newton's work on the Jewish Temple is another starting point for studying the unity of his many writings on the ancient religion of Noah and his sons around prytanaea,[24] of chronology (including profane and sacred history), prophecy, theology, alchemy, cosmogony, physics, optics and mathematics. Ancient Jewish science and its rites, like the Noachian Prytanaea, played a major role of restoration. Newton held that Jewish ancient science contained the most concise pristine knowledge of the God of dominion. It contained the hidden secrets of the order of God's design at creation and was inscribed in great detail in the proportions and measurements of the Tabernacle and the Temple of Solomon. This was the earliest knowledge regarding the working of Nature (even earlier than the science of alchemy, which was a more corrupted stage of the true science).[25] In Newton's work the ancient Jewish science, specifically the measurements of the Temple together with Daniel's and John's prophecies represent the design of God's interventions in world history. The Hebrews became the prophetic guardians of the faith given to Noah. Scriptural prophecy was granted to the Israelites in order to restore the corrupted original religion and to tell the world of the apocalyptic future.[26]

Newton's more esoteric manuscripts have a curious history. The theological and historical manuscripts were rarely discussed or scrutinized until the 1970's. In his life time Newton published only what the public could receive, ensuring that his more esoteric and heretical manuscripts remained concealed. He knew, as Steve Snobelen says, "the great damage the stain of heresy would do to the cause of his reformation in natural philosophy" if his more esoteric research were brought the public attention.[27] Only in 1936 did

---

[24] The Prytaneum was the public hall of a Greek state or city, in which a sacred fire was kept burning. Newton's historical studies pointed out that Noah and his sons spread this kind of rite around sacred fires all around the ancient world.

[25] Newton did "not find the ultimate truth [in alchemical writings], and though he appreciated the moral purpose of alchemists… only the hieroglyphs of the Biblical prophecies themselves contained Gods direct word." Manuel, *The Religion of Isaac Newton*, p. 46

[26] Goldish, *Judaism*, pp. 39-55, 64.

[27] Snobelen, "Isaac Newton, Heretic," p. 419.

his many alchemical and theological manuscripts meet with exposure. On a few occasions before that date Newton's family had tried to interest Cambridge University and the British Museum in acquiring that material but these institutions refused since the papers were outside the fields of Newton's celebrity. The option of auction remained. In 1936 Sotheby's sold off the alchemical and theological papers. Lord Keynes bought most of the alchemical material and the theological papers were bought by a wealthy Jewish Palestinian scholar, A.S. Yahuda, who was a well-known Arabist (his *The Accuracy of the Bible* had been written a few years earlier). Yahuda was an opponent of Zionism, who moved to the United States in the 1940's with his close friend Albert Einstein. Both he and Einstein tried to interest Harvard and Yale in Newton's manuscripts, but again these institutions refused, saying such documents held no interest for the scientific world. In 1951, towards the end of his life, Yahuda regretted his negative attitude towards Zionism, and decided to donate the papers to the Jewish National and University Library in Jerusalem, which did not reject the offer. The material reached Jerusalem only in the late 1960's, for family reasons.[28] David Castillejo, a Newtonian scholar, was then in Jerusalem. He arranged the documents and wrote a study of them published as a book titled *The Expanding Force in Newton's Cosmos*.[29] The book was somewhat eccentric for the taste of most of the scientific community, as will be discussed later on.

I. THE BIBLE AS A HISTORICAL SOURCE: ASTRONOMY AND HISTORY

The first group of scholars to study Newton's esoteric manuscripts on the ancient science were Castillejo, Frank Manuel, Westfall and Dobbs. In the 1980's Westfall wrote a short paper entitled "Newton's Theological Manuscripts," which describes the extent and content of these manuscripts.[30]

---

[28] For a fuller account on Yahuda, see: Richard H. Popkin, "Newton and Maimonides," Ruth Link-Salinger (ed.), *A Straight Path: Studies in Medieval Philosophy and Culture*. (Washington, D.C., 1988), pp. 216-229, 216-218; Richard S. Westfall, *Never at Rest*, pp. 875-77.

[29] Castillejo, *The Expanding Force*. There is also a short document in the Yahuda MS. written by Castillejo giving precise information on what is contained in the documents (labeled *A Report on the Yahuda Collection of Newton MSS*). There is also more information in his three volume dissertation labeled "A Theory of Shifting Relationships in Knowledge as Seen in Medieval and Modern Times with a Reconstruction of Newton's Thought and Essays on Patronage," Ph.D diss., 3 vols. (Cambridge University, 1967).

[30] Westfall "Newton's Theological Manuscripts,"; See also: idem, "Isaac Newton's *Theologiae Gentilis Origines Philosophicae*," W. Warren (ed.), *The Secular Mind: Transformations of Faith in Modern Europe- Essays Presented to Franklin L. Baumenr, Randolf W. Townsend Professor of History, Yale University* (New York, 1982), pp. 15-34; idem, *Never at Rest*; Castillejo, *Expanding Force*; Frank Manuel, *Isaac Newton: Historian* (Cambridge, 1963); idem, *The Religion of Isaac Newton*; B.J.T. Dobbs, *The Foundations of Newton's Alchemy, or 'The Hunting of the Greene Lyon'* (Cambridge, 1975); idem, *Janus Faces*; idem, "Newton as Alchemist and Theologian," pp. 129-140.

Westfall included this information in his biography of Newton - *Never at Rest*. However, he had little to say regarding the ancient Jewish science, devoting only a few words to Newton's studies of the Temple and the role of the Jewish people in the prophecies. Dobbs' pioneering work on Newton's alchemical manuscripts has attracted the attention of many scholars and has given rise to much research on his esoteric writings. Regarding Newton's attitude to ancient Jewish science, Dobbs too had little to say though she made clear the relevance and centrality of prophecy and biblical studies to Newton's alchemical and scientific work.[31]

It was mainly Frank Manuel who emphasized the central role of the ancient Israelites in Newton's historical studies. Manuel's monumental pioneering work on Newton's historical studies and religion dealt at length with his attitude towards the Bible and the Israelites in the wider European intellectual setting of the early modern period. Examining the detail of Newton's narrative, Manuel presented the method Newton used to demonstrate that the Bible was the most reliable source for ancient world history. Newton devised a chronology of kingdoms relying on astronomical data from the ancient world and comparisons between different ancient historical narratives.[32] Manuel defined Newton's approach as one of adapting data already known in astronomy (such as the precession) to the domain of historical chronology. The aim of Newton's project "was the establishment of a relationship between the observed movement of the earth with respect to the fixed stars and ancient political events, so that the past might be 'predicted' backward, so to speak."[33]

Though Newton accepted the Bible as a reliable historical source he made a distinction between the narratives and the prophetic books.[34] The narratives were not God's word but compilations of the writings of Moses, Joshua, Ezra and other records. Nevertheless, in contrast to Spinoza's radical conclusion that the Bible was no more than a historical document, for Newton scripture still remained a crucial source of God's revelation to humanity, even if many parts of it were written by human beings. Popkin continued Manuel's line of research on Newton's use of the Bible as a historical document in the context of the biblical scholarship of his time. In two informative papers, Popkin described Newton's account of how the text of the Old Testament got

---

[31] Dobbs, *Janus Faces*, pp. 73-88, 152-162.

[32] Manuel, *Newton: Historian*, pp. 65-77; 139-165; see also: Isaac Newton, *Chronolgy of Ancient Kingdoms Amended* (London, 1725).

[33] Manuel, *Newton: Historian*, p. 68.

[34] Westfall who studied "*Theologiae Gentilis Origines Philosophicae*" argues that Manuel's assertion that Biblical chronology was the touchstone through which all heathen chronologies have been tested is wrong. According to him, Newton's manuscripts reveal that Newton went over all the ancient sources and discovered traces of the "original true Religion" preserved in part by the Gentiles and in part by the Hebrews. Westfall, "Isaac Newton's '*Theologiae Gentilis Origines Philosophicae*'."

into its present state.[35] Along with Spinoza, Newton recognized the Hebrew text as having become corrupted, because of various historical events. This conclusion suited his general historical narrative of humanity's descent into idolatry. According to his astronomical calculations, the events described in the Bible took place before the earliest events recorded in Greek and Egyptian history. He concluded, therefore, that the ancient Israelites possessed the first civilization and the first monarchy, all other cultures and kingdoms being derived from the original Hebrew one.[36] Relying on his detailed compilations from other ancient cultures and astronomical data, Newton concluded that the Bible was historically accurate, however corrupted the text had become over the years. This signified to him that God had presented his message through the Israelites and through their prophets. Prophecy was a direct divine communication.

Westfall, who was the first scholar to study Newton's manuscript, "*Theologiae Gentilis Origines Philosophicae*," in detail pointed out that Newton treated ancient pagan sources as equally authoritative as the Bible. At certain places Newton gives credence to other historical sources of the ancient world over the Mosaic account, his attempt at demystification leading him to adopt such a method.[37] Westfall argues therefore that Manuel's assertion that biblical chronology was Newton's touchstone through which all heathen chronologies were tested is wrong. The "*Theologiae Gentilis*" manuscript shows that Newton scanned the ancient sources and discovered traces of the original true religion, preserved in part by the Gentiles and in part by the Hebrews.

For Newton, the prophetic books (especially Daniel and Revelation) were God's direct revelations of a hidden truth written in a symbolic and hieroglyphical language. The meaning of prophecy was central to Newton's project. He held that God delivered messages in such esoteric language because He wanted only the wise and moral to decipher them, since they were the only ones capable of properly understanding his governance in the natural world. "The scientists and divines," writes Manuel, "in Newton's entourage were not troubled by a conflict of science and theology; in their various works they were demonstrating that the world was one creation and that all its parts bespoke the same providential order."[38] The prophet, like the philosopher of nature, aimed to understand the ways of providence hidden in scripture and

---

[35] Richard H. Popkin, "Newton's Biblical Theology and his Theological Papers," P.B. Scheurer & G. Debrock (eds.), *Newton's Scientific and Philosophical Legacy*, (Dordrecht, 1988), pp. 81-98; idem, "Newton as a Bible Scholar," *Essays on the Context, Nature, and Influence of Isaac Newton's Theology*, pp. 103-118.

[36] Popkin, "Newton as a Bible Scholar," pp. 110-11. Manuel, *Newton: Historian*, pp. 89-102.

[37] Westfall, "Isaac Newton's '*Theologiae Gentilis Origines Philosophicae*'" p. 32. On Whiston's criticism on Newton's historical method, see: Force, *William Whiston: Honest Newtonian* (Cambridge, 1985), pp.140-2.

[38] Manuel, *Newton: Historian*, p. 140.

nature.[39] Newton defined the true prophet according to the writings of Maimonides, whose anti-mystical works he highly esteemed.[40] The true prophet was a supremely rational man, a man worthy of receiving a divine message, a "religious teacher who had been favoured and chosen by God because of his hard-won rational perfections, not his unbridled flights of fantasy."[41]

In his *Chronology of Ancient Kingdoms Amended* (London, 1728), Newton set out to show that the Kingdom of Israel was the first large-scale political society which had all the attributes of civilization. The Israelites were not only unique in their knowledge of the true God but also preeminent in the invention of arts and sciences. Knowledge of arts and sciences radiated to other nations from Solomon's kingdom. As Manuel points out, Newton's conception of the centrality of Judaism in the ancient world was part of a long controversy that started with the Church Fathers' insistence that the antiquity, and therefore superiority, of Moses was far beyond that of the Greeks and Egyptians. Newton's contribution to this tradition of thinking consisted in the use of comparative astronomy to deduce that the Greeks postdated Solomon by almost half a century, and that Egyptian history had been falsified and expanded.[42] He also spent much effort in calculating the future history of the Jews according to prophecy, especially their return to the promised land.

Through a comparative reading of the manuscripts, Castillejo reconstructed Newton's method of deciphering the language of prophecy and the dates of the return of the Jews.[43] Recently, Snobelen worked out Newton's prediction of the return of the Jews to Israel in the context of contemporary intellectual concerns.[44] Snobelen describes Newton's elaborate theology and

---

[39] On the unity in Newton's thought regarding prophecy and natural philosophy see: Castillejo, *Expanding Force*; Manuel, *Newton: Historian*; idem, *The Religion of Isaac Newton*; Westfall, *Never at Rest*; Dobbs, *Janus Faces*; Force, "Newton's God of Dominion," pp. 75-102; idem, "Newton's 'Sleeping Argument'"; idem, "The Nature of Newton's 'Holy Alliance," pp. 247-70; Snobelen, "God of gods"; Robert Markley, "Newton, Corruption, and the Tradition of Universal History," pp. 121-145; Scott Mandelbrote, "'A Duty of the Greatest Moment,'" pp. 281-302.

[40] On this influence, see: Manuel, *The Religion of Isaac Newton*; Popkin, "Newton and Maimonides." See also: Jose' Faur, "Newton, Maimonides, and Esoteric Knowledge," *Cross Currents: Religion & Intellectual Life*, vol. 40 (1990), pp. 526-540.

[41] Manuel, *The Religion of Isaac Newton*, p. 88, see also, pp. 66-7.

[42] Manuel, *Newton: Historian*, p. 92.

[43] Though Newton was clear that in dealing with prophecy one must not try to predict future events but only use prophecy to demonstrate the work of Providence, Castillejo, undertook to a dating drawn from Newton's work on the return of the Jews. He offers two graphs of past and future dates of historical events based on Newton's interpretation of Daniel and Revelation. According to these graphs the call of the Jews is to be in 1899 followed by 49 years recorded in the prophecy of Daniel, which means 1948. This same year - 1948 (the day State of Israel was founded) - was predicted by Newton to be the second coming of Christ. Castillejo, *Expanding Force*, p. 51.

[44] Snobelen, "'The Mystery of this Restitution of all Things': Isaac Newton on the Return of the

eschatology of the Jewish Restoration, pointing out that for Newton the return of the Jews was not only a central theme in biblical prophecy, but was also linked with all the major apocalyptic events to come. Prophecy was a divine challenge, a test to try the faithful providing, a standard by which to distinguish the future from the present, and the wise from the wicked. Newton believed that in his own time only a few could understand prophecy properly since the true time had not yet come. Very few of his own degenerate age would fathom the depths of prophecy, since "only the time of the end would realize a blossoming in such understanding."[45] As the captivity of the Jews was correctly predicted by Moses and by Isaiah, so the return would act as the ultimate evidence of the validity of biblical prophecy.[46] Corrupt prophetic hermeneutics were the cause of Israel's rejection of the person of the Messiah and a crucial reason why Christians of Newton's own time could and did fail to understand prophecy, believing that the end had already come although the Jews have not yet returned to the promised land.

II. Prophecy as a Revelation of Providence in Human History

Newton belonged to a tradition of Jewish and Christian thinkers who held that the fulfillment of prophecy in history was one of the most convincing proofs of the truth of Scripture. He believed that the prophetic books presented the history of things to come, and could be understood only after the events prophesied had actually taken place. His intention in the *Chronology* and his writings on prophecy is to give a complementary presentation of the same historical narrative of the ways of Providence in human history. James Force discusses Newton's attitude towards prophecy as a scientific proof for the presence of providence through the works of Newton's disciple William Whiston. For both Newton and Whiston the Bible was "a kind of record book of prophetic predictions communicated by God to man," and "even though we may not be able to predict exactly when the event will in fact occur, we may still be confident that the event will happen and that after it does we will be able to see how it fits in God's providential plan."[47] James Force extends this exposition to show that Newton's overall system in both prophecy and natural science had a common aim: to prove the work of providence.[48] On this issue of

---

Jews," James E. Force and Richard H. Popkin (eds.), *The Millenarian Turn: Millenarian Contexts of Science, Politics, and Everyday Anglo-American Life in the Seventeenth and Eighteenth Centuries*, (Dordrecht, 2001), pp. 95-118.

[45] Snobelen, "The Return of the Jews."

[46] On Newton's commitment to the prophecy argument, see James Force, "Newton's 'Sleeping Argument,'" pp. 109-27.

[47] Force, "Newton's Sleeping Argument," p. 121.

[48] For Newton's interpretation of the role of providence in the natural world see also: M. A. Hoskin, "Newton's Providence and the Universe of Stars," *Journal of the History of Astronomy* vol. 8 (1977),

providence, Scott Mandelbrotte adds that Newton's theology was a way "by which God could provide for his people regularity, without the need for constant acts of special providence (indulgence); in this it was similar to his natural philosophical exploration of divinity, which showed God's universe functioning according to the laws of a general providence, but preserved by specially provident mechanisms, such as the restorative actions of comets."[49]

III. THE LANGUAGE OF PROPHECY

Newton's writings on prophecy were published in *Observations upon the Prophecies of Daniel, and the Apocalypse of St. John* (1733) and appear in three major manuscripts (Yahuda MS. 1, MS. 9 and Keynes MS. 5).[50] He saw the study of scriptural prophecy as the most direct and least mediated path to understanding God's mind, since:

> The authority of emperors, kings, and princes, is human. The authority of councils, synods, bishops, and presbyters, is human. The authority of the prophets is divine, and comprehends the sum of religion, reckoning Moses and the apostles among the prophets... Their writings contain the covenant between God and his people, with instructions for keeping this covenant; and predictions to things to come."[51]

Nonetheless, he maintained that a scientific guide was needed to properly comprehend its hieroglyphic language:

> For understanding the prophecies, we are in the first place, to acquaint ourselves with the figurative language of the prophets. This language is taken from the analogy between the world natural, and an empire or kingdom considered as a world politic.[52]

---

pp. 77-101; David Kubrin, "Newton and the Cyclical Cosmic: Providence and the Mechanical Philosophy," *Journal of the History of Ideas* vol. 28 (1967), pp. 324-346.

[49] Scott Mandelbrotte "'A Duty of the Greatest Moment,'" p. 290. On comets see: Schechner, "Newton and the Ongoing Teleological Role of Comets;" Simon Schaffer, "Comets & Idols".

[50] For a detailed discussion on these manuscripts see: Castillejo, *A Report on the Yahuda Collection of Newton MSS. Bequeathed to the Jewish National and University Library at Jerusalem* (Jerusalem, 1969), pp. 2-5; Goldish, *Judaism*, 57-63, esp. pp. 60-62 and n.14 on p.60; Westfall, "Newton's Theological Manuscripts," pp. 141-3.

[51] Newton, *Observations upon the Prophecies of Holi Writ Particularly the Prophecies of Daniel and the Apocalypse of St. John*, Samuel Horsley (ed.), *Isaaci Newtoni Opera Quae Exstant Omnia* (London, 1785), vol. v, part I, chapter 1, # xvi, pp. 304-5.

[52] Newton, *Observations*, chapter 2, #i, p. 306.

Newton was influenced by Joseph Mede, "whose *Clavis Apocalyptica* (1632), introduced a synchronic calculus" and a concrete millenarian historicism, and rules for the interpretation of the prophecies of Daniel and Revelation.[53] Mede maintained that "to interpret prophecy was a grace and favour of God comparable to prophecy itself."[54] Like Mede, Newton compared prophecy with its synchronic fulfillment in history.[55] In the interpretation of the prophecies Newton used Mede's basic principles that a day in prophetic language is equal to a year in current terms, and that Revelation contains a kind of recapitulation of Daniel. Further more, an event mentioned in a fulfilled historical prophecy in Daniel as transpiring in the "world natural" refers simply to future events in the "world politic."[56]

"Newton believed," adds Goldish, "that all the prophets, indeed all Mediterranean peoples, had a standard symbolic language, which was used in the Bible as well." Through the comparative study of texts using this language in the ancient world Newton was able to construct a lexicon describing the uses of all prophetic terms "by proceeding in a universal hierarchy."[57] Mandelbrote, comparing Newton's lexicon with the work of other figures in the Mede school, adds that "in general [Newton's] technique is crude, but its hall mark is consistency and it was this that made it a viable and desirable tool of interpretation."[58] Castillejo notes Newton's tendency to interpret prophetic symbolism in political terms and devises tables and figures that reflect Newton's interpretation of the prophecies.[59] The following is an example of Newton's lexicon:

> The whole world natural consisting of heaven and earth signifies the whole world politique consisting of thrones and people. ... When a Man is taken in a mystical sense his qualities are often signified by his actions & the circumstances of things about him. So a Ruler is signified by his riding a Beast...[60]

---

[53] Goldish, *Judaism*, pp. 24, 57.

[54] Manuel, *The Religion of Isaac Newton*, p. 91; Castillejo goes as far as saying that "Newton's work is avowedly and consciously modeled on Mede's," *A Report*, p. 4.

[55] On the relation between Mede and Newton, see: Sarah Hutton, "More, Newton, and the Language of Biblical Prophecy," *The Books of Nature and Scripture*, pp. 39-53; Sarah Hutton, "The Seven Trumpets and Seven Vials: Apocalypticism and Christology in Newton's Theological Writings," *Newton and Religion*, pp. 165-178.

[56] On these issues, see: James Force, *William Whiston: Honest Newtonian* (Cambridge, 1985), pp. 72-3.

[57] Goldish, *Judaism*, p. 61

[58] Mandelbrote, "Biblical Criticism," p. 297.

[59] Castillejo, *Expanding Force*, pp. 32, 35.

[60] Newton, KMS. 5, cited in Castillejo, *Expanding Force*, p. 32; see also: Newton, *Observations*,

The meaning of the prophecy and its relation to the worship around the Tabernacle and the two Jewish Temples is discussed at length in chapter nine of this book. For the time being it is enough to note that from these prophecies Newton devised a mathematical formula describing the historical process of the corruption of the Israelites through the expanding dimensions of the physical structures of the Tabernacle and the two Jewish Temples.

IV. NEWTON'S ROLE IN THE RESTORATION OF THE ORIGINAL RELIGION

Newton gives a clear indication of how he saw his role in re-instilling a lost Truth:

> Having searched & by the grace of God obteined knowledge in ye prophetique scriptures, I have thought my self bound to communicate it for the benefit of others, remembering ye judgment of him who hid his talent in a napkin.[61]

Many scholars have pointed out that Newton's natural philosophical method (of underlying simplicity) is similar to the "way he organizes other areas of inquiry."[62] For Newton truth was always simple whatever its subject matter. The wise of each generation have a duty to uncover the simplicity governing all seemingly complex phenomena, whether the motion of bodies, the constitution of rays of light, alchemical cycles, the words of prophecy, or the work of providence in history. All these areas of inquiry exhibit the original ordered design God bestowed on the world at creation. Indeed, Kenneth Knoespel argues that Newton's unfinished manuscript of the *Theologiae Gentilis Origines Philosophicae* was to be "devoted to showing how ancient religious practice could reveal physical truths about the universe."[63] Similarly, Mandelbrote writes that Newton's work on prophecy assumes that "mathematics was God's language."[64] James Force even declares that "Newton's calculus is based on the continuity of flow as supervised by the God of Dominion operating in his generally provident mode of creator and preserver of the current state of natural law."[65] These arguments for the common scientific aspects of Newton's diverse interests are reinforced by Niccollo Guicciardini's research on Newton's use of mathematics in the

---

chapter 2, "Of the Prophetic Language."

[61] Newton, Yahuda MS. 1.1, fol.1r, in Sarah Hutton, "The Seven Trumpets," p. 167.

[62] Hutton, "The Seven Trumpets," p. 165. On this issue see also: Maurizio Mamiani, "The Rhetoric of Certainty: Newton's Method in Science and in the Interpretation of the Apocalypse," M. Pera and W.R. Shea (eds.), *Persuading Science: The Art of Scientific Rhetoric*, (Canton, 1991), pp. 157-172; Force, "Newton's God of Dominion"; Snobelen, "God of gods"; Markley, "Newton. Corruption and the Tradition of Universal History."

[63] Knoespel, "Interpretive Strategies," pp. 179-203, 180.

[64] Mandelbrote, "A Duty of the Greatest Moment," p. 301.

[65] Force, "Newton's God of Dominion", p. 88.

Principia. Guicciardini points out that Newton made a "methodological turn of the 1670's" regarding the method of fluxions. In this shift Newton thought to recover the geometrical procedures of the ancients, and incorporated them in his new physics. This shift fits into Newton's overall reorientation toward the *prisca sapientia* and his role in reviving the lost truth.[66]

In what follows I will consider the purpose and ideas of this new religion of natural philosophy, revealed "by grace" to Isaac Newton. More specifically, I will discuss Newton's understanding of the mathematical principles conditioning God's original design for the world and the mathematical heavenly order imprinted upon matter in creation and sustained from day to day. I shall argue that Newton's method of fluxions had a theological basis. First I shall present Newton's mathematical theory and its definitions and axiomatic choices. Then I will describe how this mathematical method is incorporated into the definitions and laws of motion in the *Principia*. Next, I will analyze his notions of space and time and of their connection to God's unconditional creative source of energy, going on to the separate writings on cosmogony, world history and the work of providence. Following, I will discuss the role of alchemy in Newton's decipherment of the spiritual secrets behind the ordered design, and his approach to it as a science of the wise to assist God in amending and fixing what has become materially corrupted. I will conclude with an analysis of Newton's studies on the prophecies of Daniel and Revelation and the Jewish Tabernacle and the two Temples, which to my mind are his clearest representations of the mathematical nature of the laws of human corruption.

Put differently, the book starts with the method of fluxions and moves on to physics in order to give more content to the mathematical conditions conferring reality on the original design. From the physical practical exposition of the design I go on to the articulated historical plan of providence as it appears in Newton's work on prophecy, profane and sacred history, and cosmogony. Finally, the book studies Newton's research in alchemy, prophecy, and the Jewish Temple, which concretizes the world design in a mathematical cryptogram showing how human beings who know the true science can assist God in restoring what has become corrupt. All these studies converge the original order of the design that ought to be persevered, and how misuse or mishandling through idolatry have obstructed and almost completely destroyed it. This is only one side of the story; prophecy reveals a still more meaningful aspect, namely, that a hidden plan of providence exists. God replenishes and amends through comets what has become corrupted in the natural world, and through spiritual messengers what has been damaged in human history, thus periodically revealing his central role as the Governor and Lord and Master of creation.

In chapter II, I present Newton's unified and systematic project as guided by the notion that God has imprinted upon matter a simple ordered

---

[66] Niccollo Guicciardini, *Reading the* **Principia**: *The Debate on Newton's Mathematical Methods for Natural Philosophy from 1687-1736*, (Cambridge, 1999), pp. 37-38, 99-104.

design with symmetrical features. This order has a mathematical structure also in the sense that the laws of motion of the *Principia* (especially the third law governing material actions and reactions) have a clear mathematical aspect, as his title suggests – *The Mathematical Principles of Natural Philosophy*. These general laws dealing with the outcome of material interactions condition God's design, and as I shall show in the following chapters, they apply also to the history of humanity once read as general laws of actions. More specifically, the axiomatic part of the method of fluxions, which Newton developed quite early in his life, contains the secret of the operation of the design. It tells how God sustains creation and which actions preserve and keep the design functioning according to the original plan, and which corrupt it and become destructive, forcing God to intervene at certain periods in history. This reading of the method of fluxions enables us to understand the mechanism of the corruption through idolatry of the pristine Noachide religion in Newton's thought.

Chapter III points out that Newton's youthful method of fluxions (1660's), which later went on to what Guicciardini calls "*prisca geometrica,*" contains a seed analogous to the notion of God's original pure Vestal religion. The concepts behind the method of fluxions (mainly the equable flow of time) and the later shift are a mathematical interpretation of the root of corruption in Nature and society and an abstraction of the lawfulness governing material actions. Newton's method of fluxions and his later geometrical methods in the *Principia* (the recovery of the *prisca geometrica*) seemed to demonstrate to him a fundamental truth about the decline of all processes in nature. Indeed, according to Markley, for Newton, "mathematics becomes akin to revelation," and is the "analytical and moral shield against the corruption of the physical world."[67] The knowledge contained in the method of fluxions opens a path for the human mind to follow in understanding how to operate God's worldly design and thus worship him more correctly.

In order to appreciate the particularity of Newton's method of fluxions I compare his mathematical and physical insights to those of Leibniz, his contemporary, who developed the calculus separately, together with a sophisticated scientific and metaphysical system. Chapter III analyzes the method of fluxions from a philosophical point of view, pointing out its epistemological assumptions. Chapter IV is devoted to a discussion of Leibniz's calculus, dwelling on the metaphysical insights contained in his mathematical analysis. Neither of these chapters requires a strong mathematical background from the reader. They are written in such a way that the intuitions of the two respective mathematicians may be grasped together with their theological implications. Chapter V elaborates further on these respective mathematical intuitions, associating them with the two opposing notions of space and time. Much has been written on this subject; what I have tried to add to is the spiritual application and the inherent connection between the mathematical and physical concepts and God's relation to his creation. Chapters VI and VII discuss Newton's and Leibniz's opposed notions of God's

---

[67] Markley, "Newton, Corruption, and the Tradition of Universal History," p. 140.

infinite perspective, remarking that their respective mathematics function as an epistemological purifying tool intended to bring the student closer to the divine point of view. Chapter VI, dealing with Newton's notion of God's absolutes, also touches upon the structural analogies apparent in his many writings on cosmology, prophecy, history, physics and the method of fluxions. Chapter VII points out the similar structures appearing in Leibniz's metaphysical system of monads, his laws governing monadic perceptions, his laws of the motion of bodies, and his calculus. The chapter ends with a comparative discussion of the two respective mathematics and the ways in which they may be seen too function as spiritual tools for achieving the epistemological purification necessary for reaching God's perspective.

Chapter VIII contains a more detailed discussion of Newton's writing on the Vestal religion and its structural resemblance with his method of fluxions. It also argues that his writings on alchemy may give us a better understanding of his belief in the secrets he thought to lie behind the ceremony of the sacrificial fire and behind all cyclical phenomena in creation. In chapter IX I analyze Newton's work on the Tabernacle and the Temples from the point of view of the mathematical components of the design of the world and the laws of action governing it. Chapter X attempts to draw together the various aspects of Newton's system and see how they fit together.

Throughout the book I employ two levels of discussion. On the first level I present Newton's many dispersed ideas on the order of the designed world, remaining as close to his original ideas as possible. In the discussion on the second level I try to use my own synthetic abilities, bringing in modern scientific theories that may shed light on Newton's system. As the reader will see I employ current notions about built-in human design in computers and other modern tools in connection with Newton's ideas on God's design.

CHAPTER II

THE MATHEMATICAL PRINCIPLES OF GOD'S DESIGN

"For God is known from his works."[1]
"In the preceding Books [I and II] I have laid down the principles of philosophy, principles not philosophical, but mathematical." (*Principia*, p. 319.)

Newton was attempting to discover the original design of creation imprinted upon matter, for as he said, "God is known from his works."[2] Two major routes were open to him in following up the mathematical principles of this original design. The first was the natural world; the second was the Bible (mainly the prophetic books) and the works of providence in the history of humanity. In both cases his working assumption was that a real design exists in nature and in human history, which the scientist and the interpreter of prophecy have a duty to uncover in order to provide human beings with practical guidelines for preserving and taking better care of the order of the design. Even prior to the events of the Glorious Revolution (1688-9), natural philosophers and divines of the Royal Society such as John Wilkins and Robert Boyle had emphasized the providential role of God in first ordering and then supervising the course of nature.[3] In the context of the renewed political instability arising from the Glorious Revolution the providential Newtonian design argument was further popularized by scientists-theologians, such as Richard Bentley, Samuel Clarke and William Whiston in the Boyle Lectures, which were supervised by Newton himself. According to Westfall, these men along with Edmond Halley and Hopton Haines shared Newton's heretic Arianism which had a direct

---

[1] J.E. McGuire, "Newton on Place, Time, and God: An Unpublished Source," *BJHS* vol. 11, (1978) p.119. See also: "We are, therefore, to acknowledge one God, infinite, eternal, omnipresent, omniscient, omnipotent, the Creator of all things, most wise, most just, most good, most holy. We must love him, fear him, honor him, give him thanks, praise him, hallow his name, obey his commandments, and set times apart for his service, as we are directed in the Third and Fourth Commandments.... And these things we must do not to any mediators between him and us, but to him alone... And this is the first and the principal part of religion. This always was and always will be the religion of all God's people, from the beginning to the end of the world." Newton, "A Short Scheme on the True Religion," Quoted in Brewster, *Memoirs of Sir Isaac Newton*, (Edinburgh, 1855), vol. 2, p. 348.

[2] J.E. McGuire, "Newton on Place, Time, and God: An Unpublished Source," *The British Journal for History of Science*, vol. 11, (1978), p. 119.

[3] For an account on the revealing of the providential design by members of the Royal Society, see: James Force, "Hume and the Relation of Science to Religion among Certain Members of the Royal Society," *Journal of the History of Ideas*, vol. 45 (1984); idem, *Whiston*, pp. 90-95; Margaret C. Jacob, *The Newtonian in the English Revolution: 1689-1720*, (Ithaca, 1976).

saying on the design argument.[4] Margaret C. Jacob points out that these Boyle Lectures were used to spread the design argument of Newtonian natural religion to the social needs of Restoration England.[5] Indeed, Newton informs Richard Bentley before Bentley's first Boyle lecture that:

> when I wrote my treatise about our Systeme I had an eye upon such Principles as might work with considering men for the beliefe of a Deity, and nothing can rejoice me more than to find it useful for that purpose.[6]

The idea of an original design that has mathematical and geometrical qualities is central in Newton's mind. In both his major scientific publications, the *Principia* and the *Opticks*, the "argument from design" provides the key theological metaphor of the world as a system of God's "most wise and excellent contrivances."[7] Although by the seventeenth century the argument was serving as the conventional metaphysical means for demonstrating God's existence, Newton's endeavors to bring the metaphor it conveyed to bear directly on the philosophical study of nature and its underlying mathematical principles were not commonplace. Natural philosophers traditionally sought to explicate fundamental natural properties, that is, the simplest forms of nature underlying the phenomena. These explications provided the premises on the basis of which the phenomena could ideally be demonstrated. Thus, the method of "resolution" guided inquiry from evidence to basic premises, while the method of "composition" was followed in the attempt to investigate the consequences that followed the premises of inquiry.[8] Although it was generally assumed that simple natures were created by God's omnipotence and omniscience, the logical structure of demonstrations did not pose the practical challenge of understanding a work of design. The latter would involve the examination of functional relations between the various parts that are

---

[4] See: Westfall, *Never at Rest*, p. 651; Force, *Whiston*, p. 93.

[5] Margaret C. Jacob, *The Newtonian in the English Revolution*, p. 270; See also: Force, *Whiston*, p. 63.

[6] "Newton to Bentley, December 10, 1692", W. H. Turnbull, J.F. Scott, A. Rupert Hall and Laura Tilling (eds.), *The Correspondence of Isaac Newton* (Cambridge, 1959-77), vol. 3, p. 233.

[7] Newton, *Isaac Newton, The Principia: Mathematical Principles of Natural Philosophy, a New Translation by I. Bernard Cohen and Anne Whitman, Assisted by Julia Budenz.* Cohen and Whitman (eds.), (Berkeley: University of California Press, 1999), p. 942. I wrote parts of this book throughout the years before this wonderful translation and edition was published, so I only cite from this edition (hereafter cited as *Isaac Newton, The Principia*) and not from Andrew Motte's translation when I find it necessary. Cohen also demonstrated that natural theology was explicit in all three editions of the *Principia*, see: I.B. Cohen, "Isaac Newton's *Principia*, the Scriptures, and the Divine Providence," Sidney Morgenbesser, et al. (ed.), *Philosophy, Science and Medicine* (New York, 1969), pp. 523-48.

[8] See e.g.: A.C. Crombie, *Robert Groseteste and the Origins of Experimental Science, 1100-1700* (Oxford, 1953), esp. pp. 290-319.

integrated by the order of the design into a particular synthesis that a logical inference cannot fully disclose or recover.[9]

It is important to keep in mind in this context the difference between grasping nature as God's *creation* and grasping the world as a work of God's *design.* Alluding to this distinction, Newton pointed out in his preface to the *Principia* that "[t]he ancients considered mechanics in a two-fold respect as rational, which proceeds accurately by demonstration; and practical."[10] "Rational" mechanics sought to disclose the logical relations between specific magnitudes that a particular device must logically involve working. Artisans, on the other hand, faced the practical challenge of making a particular device, of learning the contingent constraints that it presented and the opportunities it offered in virtue of its special design. Rational mechanics did not provide the guidelines for accommodating the user's conduct to the designed system, while the user's practical experience fell short of demonstrative knowledge. As a philosopher, Newton was committed to the search for methodologically rigorous explanations; yet, because he construed the world as a work of design, he concluded that practical experience was the indispensable element in the acquisition of this knowledge. "Accurate practice" thus appeared to be the appropriate ideal: "he that works with less accuracy is an imperfect mechanic; and if any could work with perfect accuracy, he would be the most perfect mechanic of all."[11]

In the 1710's, Samuel Clarke on behalf of Newton further developed the argument of design. In the famous correspondence between Leibniz and Clarke, Leibniz criticizes Newton's concept of design as follows:

> According to [Newton's] doctrine, God Almighty wants [i.e. needs] to wind up his watch from time to time; otherwise it would cease to move. He had not it seems, sufficient foresight to make it a perpetual motion.... According to my opinion, the same force and vigor remains always in the world and passes from one part of matter to another, agreeably to the laws of nature and the beautiful pre-established order.[12]

Here Leibniz accuses Newton of detracting from the power and majesty of God by envisioning his creation as imperfect, whereas the Leibnizian God created from the beginning the most perfect design that develops according to determined laws of nature. In other words, while Newton's God has to intervene to keep the design of the world on track,

---

[9] I owe these insights and the discussion in the following paragraph to Michael Ben-Chaim.

[10] Isaac Newton, *The Principia: translated by Andrew Motte*, (New York, 1995), p. 3. Hereafter cited as *Principia*.

[11] Newton, *Principia*, p. 3.

[12] Leibniz's first reply, Loemker, Leroy E. (ed.), *G.W. Leibniz: Philosophical Papers and Letters* (Dordrecht, 1989), pp. 675-6.

Leibniz's God created the universe according to such perfect mathematical and logical laws that he has no need to intervene. Once the axioms or laws of nature were chosen the best initial conditions were also chosen (since the calculus determines the best choice).[13] Therefore God no longer needs to intervene. Newton's answer, conveyed by Clarke, is noteworthy because he claims that God's constant intervention in the world is no indication of weakness but on the contrary, a sign of his enormous power:

> … the skill of all human artificers consists only in composing, adjusting, or putting together certain movements, the principles of whose motion are altogether independent upon the artificer. Such are weights and springs and the like, who are not made but only adjusted by the workman. But with regard to God the case is quite different, because he not only composes or puts things together, but is himself the author and the continual preserver of their original forces or moving bodies. And consequently 'tis not a diminution but the true glory of his workmanship that nothing is done without his continual government and inspection.[14]

Let us look more closely at Newton's understanding of God's design of the universe and his role in this design. In such a design God must employ non-mathematical elements, otherwise, as Leibniz thought, everything would be determined according to the mathematical laws of nature. However, to Newton, without these non-mathematical features God could have never been called the "governor" of the world. God purposely chose not to create a perfect mathematical design at creation because he was a "hands on" God and intended to intervene in historical events. God is consequently "the continual preserver" of the original design throughout history. God designs creation and sustains it on a daily basis.[15] Newton could not accept a creation that restricted and limited God's will and did not allow room for human error.[16] God created the world in such a way that it could become corrupted and damaged as a result of human actions, otherwise there would be no need for his intervention. This is precisely what Newton has Clarke tell Leibniz. Assuming the whole world is a machine designed in such a way, Clarke says, God will want to amend it here and there, otherwise he will only be the creator but not the governor of the design.[17] "The notion of the world's being a great machine," writes Clarke,

---

[13] This issue is discussed at length in the chapter on Leibniz's calculus.

[14] Clarke's first reply, *G.W. Leibniz: Philosophical Papers and Letters*, pp. 676-7.

[15] The Jews state in their daily prayers, a similar idea, that God "in his goodness creates anew every day constantly the work of Creation." This passage occurs in the penultimate blessing before the recital of the *Shema'*. On Newton's acquaintance with this pray, see: Goldish, *Judaism*, p. 160.

[16] The possibility of idolatry is necessary for the intervention of providence in history.

[17] Clarke's second reply #11, *G.W. Leibniz: Philosophical Papers and Letters*, p. 681.

"going without the assistance of a clockmaker... tends to exclude providence and God's government in reality out of the world."[18]

In the design of the material world there is a clear mathematical component. Newton tries to understand and study this mathematical design through his experimental method. But he knows that the mathematical features of the design are not its only component: if the design were wholly mathematical there would be no place for divine providence or human action. Without direct revelation man could never have discovered and understood these non-mathematical features. It is only through the revelations of chosen souls that the cultural, theological, and historical components of the design are disclosed to humanity. In the ancient world the scientific components were revealed to prophets such as Moses and Ezekiel who were instructed to encode the mathematical secrets of the design in the measurements of the Tabernacle and the Temple. From this perspective, Newton believed that prophecy was a source of truth no less than science, and that, like science, it revealed to men a mathematical lawfulness guiding the original divine design so that they could become "perfect mechanics," so to speak, also in terms of world history. I propose, therefore, that we read Newton's historical work in *Chronology* and prophecy together with his experiments in the *Principia* and *Opticks* as an exposition of the practical results of the actions and effects of this design, and of their mathematical nature as it appears in both the human and material realms.[19]

It is my understanding that Newton distinguished between the laws conditioning the actions and effects of the design and the design itself. In the correspondence, Clarke says explicitly that what might seem accidental to human beings is not so from God's perspective:

> The present frame of the solar system, for instance, according to the present laws of motion, will in time fall into confusion and perhaps, after that, will be amended or put into a new form. But this amendment is only relative with regard to our conceptions. In reality with regard to God, the present frame, and the consequent disorder, and the following renovation are all equally parts of the design framed in God's original perfect idea. 'Tis in the frame of the world, as in the frame of man's body; the wisdom of God does not consist in making the present frame of either of them eternal but to last so long as he

[18] Clarke's first reply, *G.W. Leibniz: Philosophical Papers and Letters*, p. 677.

[19] Dobbs suggests a similar correspondence in Newton's work regarding the role of spirit in both the human and natural domains: "The natural and moral worlds found for him their deepest relationship not in the details of their parallel movements but in the fact that both were theaters for the activity of the spirit. Prophecy fulfilled and correctly interpreted, provided for Newton exactly the same sort of evidence for the divine governance of the world as his "active" alchemical processes, and it served the same metaphysical purpose. There is also reason to suppose he considered the divine agent in both worlds to be the same." Dobbs, *Janus Faces*, p. 80.

thought fit.[20]

What may seem arbitrary and coincidental to men since it seems not to follow mathematical laws, nonetheless has a divine reasoning behind it, which men can only come to know if God chooses to reveal the secret to the wise.[21] In the following chapters I will discuss in greater detail Newton's insistence that the principles governing the material actions and effects of this design, in contrast to the human non-mathematical features of the design, have a clear mathematical lawfulness. The reason behind this lawfulness seems to be the general law of action defined in the material realm as the third law of motion and in the human domain as the divine law of reward and punishment. This general law that assumes that for every action that takes place within the designed universe there is a specific concrete reaction governed by mathematical laws.[22] In physics this reaction can be calculated mathematically. In the human world, however, the mathematics is too complex to calculate, yet it is clear that for every action there is a consequence according to the general laws of action. Actions that follow the design according to its original purpose keep the universe on its lawful course, while actions that do not follow it corrupt and damage the order of the design. Any science that exposes the mathematical principles of God's original design assists human beings in becoming more aware of their duty towards God and the order of the design he gave people to keep and preserve. To give an analogy: once a person understands exactly how a machine operates and what effects each of his actions will have on the machine, he will make sure to operate the machine correctly according to its design and not mishandle it. Moreover, if he does make a mistake due to carelessness or any other reason he will know how to repair the machine in accordance with its original design.

Therefore, I argue that Newton understood the commandments given to humanity through chosen messengers, such as Noah, Moses, and Christ, as being the simplest and most general principles of action that serve God in preserving the universe, in the sense that these sorts of actions do not obstruct and interfere with the worldly design. Newton is very clear on this subject. The difference between idolatrous actions and religious ones is that in true religious actions we adore and love God, "and these things we must do not to any mediators between him and us, but to him alone."[23] The commandments, which were specially designed for men, function like the axiomatic part of the instructions of a huge manual on how to operate and preserve the ordered world given to men. The bodies of human beings are also part of the design

---

[20] Clarke's second reply #8, *G.W. Leibniz: Philosophical Papers and Letters*, p. 681.

[21] Snobelen, "Isaac Newton, Heretic," pp. 418-9.

[22] In mechanics see: Newton, *Principia*, Third Law of Motion, pp. 19-20; and in the human domain a clear statement appears in the writings of Moses: Deuteronomy, VIII, 1-20.

[23] Newton, "A Short Scheme on the True Religion," Quoted in Brewster, *Memoirs of Sir Isaac Newton*, vol. 2, p. 348.

and the actions they perform with their bodies are crucial to the operation of the designed world. Thus when people stop obeying his commandments they are punished:

> As often as mankind has swerved from [the two first commandments – love of God and man] God has made a reformation. When ye sons of Adam erred & the thoughts of their heart became evil continually God selected Noah to people a new world & when ye posterity of Noah transgressed & began to invoke dead men God selected Abraham & his posterity & when they transgressed in Egypt God reformed them by Moses & when they relapsed to idolatry & immortality, God sent prophets to reform them and punish them...[24]

As this text shows, Newton's theological and historical writings set out to describe the consequences of mistaken actions upon the history of humanity. His belief that the world will be adversely affected if men practice an idolatrous religion was in conformity with the belief of the vast majority of religious people in the seventeenth century, who maintained that what people believed affected both themselves and the world at large.[25] Indeed, any one who devoted so much time and energy to the study of the Jewish Temple rites, as Newton did, would have encountered the same belief in the ancient texts. As Raphael Patai points out:

> the functioning of the natural order was believed to have depended not solely on the regular and precise performance of the Temple ritual, but also on the conduct of the people, particularly of its leaders, outside the sacred precincts. There existed, it was believed, an inner, a sympathetic connection between human conduct and the behavior of natural forces. Sins, i.e. improper or illicit human acts, brought about improper or illicit occurrences in nature.[26]

Taking into account Newton's systematic way of thinking and his

---

[24] Newton, Keynes MS. 3, pp. 35-6, also cited in Goldish, *Judaism*, p. 64.

[25] "It was a common assumption in the 17th century," writes Dobbs "that the physical state of our world paralleled the moral history of mankind." Dobbs, *Janus Faces,* p. 232.

[26] Raphael Patai, *Man and Temple: In Ancient Jewish Myth and Ritual* (New York, 1947), pp. 221-2. The Jewish Midrash Tanhuma gives us a succinct expression of the connection between human beings' body, the world, and the Temple as it was believed back then: "The Temple," it says, "corresponds to the whole world and to the creation of man who is a small world." *Midrash Tanhuma*, Pequde, & 3, cited in Patai, *Man and Temple*, p. 116. In another Jewish legend it is said that God told Moses that "just as I created the world and your body, even so you will you make the Tabernacle." Albeck, (ed.), *Bereshith Rabbati*, p. 32, cited in Patai, *Man and Temple*, p. 115.

interest in ancient ritual forms, it is difficult to resist the conclusion that he believed human actions could disrupt and damage the material world. The Israelite belief that the physical structure of the Temple corresponded to the world and to the organs of the human body[27] may underlie Newton's insistence that any form of idolatry corrupted humanity and the worldly design. Idolatry was the name Newton gave to all those actions that did not obey the intentions of the commandments. All forms of idolatry have a common feature. At certain moments in history human beings stopped worshipping God correctly and instead of being attuned to the only true source of the design they began to rely on false gods, beliefs, or ideologies.[28] The consequence of such behavior on the functioning of the natural world is further exemplified in the meaning the Israelites gave to the Temple rites:

> [T]he Temple symbolically represented the entire universe, and each and every rite performed in it affected that part or aspect of nature of which the rite itself was reminiscent. In fact the function of the Temple was of such basic importance that the very existence of the entire world depended on it.[29]

Newton's examination of the ancient religious practice of the Prytanaea also shows that in his view God designed the original religion in accordance with the mathematical features of the orderly design with which he constantly sustains the solar system. As mentioned previously, the principal ancient rite that served to communicate the mathematical knowledge of God's original design evolved around the sacrificial fire of the Prytanaea. The Prytanaea was an ancient religious practice deriving from Noah and his sons, that was celebrated later on in Vestal Temples, where there were circles surrounding a sacred fire, which in Newton's view, preserved and represented the divine wisdom of heliocentricity.[30] This ancient ceremony of priests circumambulating the sacred fire represented the true mathematical relationship of the solar system. By performing this rite correctly, the priests sustained the mathematical order of the universe; should they fail in their duties the order of the world would be damaged and in need of replenishment or amendment.[31] In his work on prophecy, Newton alludes to the analogy

---

[27] See e.g. cited in Patai, *Man and Temple*, p. 114-7.

[28] See e.g. the last paragraph in Newton's *Opticks*, pp. 405-6 and Yahuda MS. 41, p. 8r.

[29] Patai, *Man and Temple*, p. 221.

[30] For a precise and short account of the history of the original religion and the prytanaea, see: R.S. Westfall, "Newton's Theological Manuscripts," pp. 129-43. A longer account on Newton's historical studies appears in: Kenneth Knoespel, "Interpretive Strategies in Newton's *Theologiae Gentilis Origines Philosophiae*," *Newton and Religion*, pp. 193-4; Robert Iliffe, "Is He Like Other Men?" pp. 165-167; Force, "Samuel Clarke's Four Categories of Deism," pp. 56-67; idem, "Newton, the "Ancients," and the "Moderns," *Newton and Religion*, pp. 237-59, 253-4.

[31] In this religious context, the idea of using comets to replenish the original design of the solar

between the "world natural, and an empire or kingdom considered as a world politic" in the figurative language of the prophets. In the prophetic language, whatever happens in the heavens or on Earth has an analogue in the world politic. Stability and prosperity in the world politic go together with a steady and fertile nature, whereas the analogue of corruption in the human dimension is an upheaval in nature:

> Accordingly, the whole world natural consisting of Heaven and Earth, signifies the whole world politic, consisting of Thrones and People; or of so as it is considered in the prophecy: and the things in that world signify the analogous things in this.... Great earthquakes, and the shaking of heaven and earth, for the shaking of dominions, so as to distract or overthrow them.[32]

Therefore, I maintain that Newton understood creation to be a most sophisticated and wonderful gift given to men. God, as the governor of this design, gave his children a degree of freedom and responsibility in operating and sustaining it. To assist him in the operation He gave them the commandments, prophecy, and the ancient science. As long as people remained wise, obedient, moral, and spiritual (attuned only to God) they were able to enjoy the design.[33] Yet as history has shown, it is inherent in human nature that as time passes men will gradually fall from their pristine wisdom and spirituality and become corrupt as a result of material temptations. Whenever this occurs they become idolatrous. The consequence of idolatry is that men lose their spiritual freedom and become attached to material substitutes:

> Now the corruption of this religion [Prytanaea] I take to have been after this manner. First the frame of ye heavens consisting of Sun Moon & Stars being represented in the Prytanaea as ye real temple of Deity men were led by degrees to pay a veneration to these sensible objects & began at length to worship them as visible seats of divinity.... For 'tis agreed that Idolatry began in ye worship of the heavenly bodies and elements.[34]

Once idolatry infiltrates, people are no longer able to fulfill their

---

system would become necessary only if a great loss of motion has occurred due to material interactions that disrupt the daily sustenance of the original harmonic mathematical design. On the replenishment through comets, see: Schechner, "Newton and the Ongoing Teleological Role of Comets," pp. 299-311; Schaffer, "Comets & Idols," pp. 206-231.

[32] Newton, *Observations*, chapter 2, p. 306.

[33] As Moses had promised the Israelites, see: Deuteronomy, VIII, 1-20; XI, 13-27; XXX, 1.

[34] Newton, Yahuda MS. 41, p. 8r, cited in Goldish, *Judaism*, p. 51.

responsibilities in maintaining the design of the world. Even then God keeps sustaining the order of the design daily, as He did before the fall into idolatry, yet those parts for which human beings are responsible become corrupt. God's governance enters at those critical moments when the whole design is on the verge of collapse. He then replenishes the solar system with comets and reveals the true religion through messenger souls.

This way of understanding Newton's system shows why he insisted so firmly that idolatry led to the corruption of the world's design and the consequent corruption of the natural world itself, whereas true religion restores the world to its intended order. To reiterate the machine analogy from the Clarke-Leibniz correspondence: a machine will function smoothly with hardly any friction or disruption from its surroundings when it operates according to its intended design, yet it can gradually become less functional as a result of internal and external disruptions caused by misuse, until at a certain point it may almost cease to function. At this point the machine will need its designer to repair it. I maintain that Newton saw the world in a similar way, as a huge designed machine that functioned efficiently as long as humans operated it according to the original design given to them by the governor of the machine. Since human beings are both a part of this design and play a role in preserving it, their actions are of primary importance. When their actions accord with the divine commandments, everything goes smoothly; when their actions ignore or defy the commandments, they are catastrophic repercussions for both the human and material realms. As Clarke tells Leibniz, the amendment of the design is necessary precisely because so few people understand the profound secrets behind its operation.

True worship centers on God's two simple commandments. Metaphysical speculations are idolatrous when they divert believers from piously following these commandments.[35] By Newton's time the corruption of society through metaphysics was such that the ancient scientific truth represented and sustained by the priests as they circled the sacred fire could no longer be encoded in a religious ritual. Neither could Moses' description of the seven days of creation satisfy the scientific mind.[36] In a letter to Thomas Burnet in 1681, Newton explains Moses' motive in writing about creation in such a simplistic manner:

> As to Moses, I do not think his description of the creation either philosophical or feigned, but that he described realities in a language artificially adapted to the sense of the vulgar.... His business being, not to correct the vulgar notions in matters philosophical, but to adapt a description of the creation as handsomely as he could to the sense and capacity of the vulgar.[37]

---

[35] Goldish, *Judaism*, p.11.

[36] On this see: Dobbs, *Janus Faces*, pp. 54-62.

[37] The letter is in Brewster, *Memoirs of the Life*, vol. 2, pp. 447-54 also in *Correspondence*,

By Newton's time, Moses' narrative was no longer suited even to the majority of the vulgar. Something different was needed to glorify the work of God. The new scientific worldview of a heliocentric universe needed a new theological framework to support it. Newton understood that if he wanted his contemporaries to return to the commandments of the original simple religion (which were also, he thought, the pillars of Christianity) he had to find a different way to express its scientific truth. As Thomas Aquinas accommodated Greek science into Christianity, providing his generation with a coherent scientific and theological worldview, so did Newton conclude that he needed to compensate his contemporaries for the lost of meaning and integrity that the new heliocentric universe had brought to the Christian faith. Thus he undertook to present the mathematical truth of the solar system in an explicit, practical, and detailed scientific method that also corroborated the truth of the original religion.[38] Indeed, this is what Newton set out to do in Book Three of the *Principia*. He depicts the mathematical structure of the solar system, as already encoded in the worship of Noah and his sons around the sacrificial fire and in the measurements of the Tabernacle and the Temples. From this perspective, an analogy can be made between the roles of Newton and Moses respectively in world history. Premised on Moses' law, the ancient religion of the Israelites comprised a detailed set of disciplines that thoroughly mapped and structured everyday life. This system minutely specified constraints and opportunities for action and described the consequences in a wide variety of practical contexts. Analogously, Newton's studies in optics and mechanics aimed at mapping the detailed mathematical and mechanical laws that governed the universe. Thus once humans genuinely understood the design of the natural world they would return to the true ancient religion, a point he makes at the end of the *Opticks*, where he draws an explicit connection between good natural philosophy and good religion:

> For so far as we can know by natural Philosophy what is the first Cause, what Power he has over us, and what Benefits we receive from him, so far our Duty towards him, as well as towards one another, will appear to us by the Light of Nature. And no doubt, if the Worship of false Gods had not blinded the Heathen, their moral Philosophy would have gone farther than to the four Cardinal Virtues; and instead of teaching the Transmigration of Souls, and to worship the Sun and Moon, and dead Heroes, they would have taught us to worship our true Author and Benefactor, as their Ancestors did under the Government of *Noah* and his Sons before they corrupted themselves.[39]

"Newton to Burnet, Jan 1680/1," vol. 2, pp. 329-334, 331.

[38] Steve Snobelen shows that the General Scholium of the *Principia* contains Newton's theology. Snobelen, "God of gods."

[39] Newton, *Opticks*, pp. 405-6.

Like Moses' commandments, Newton's experimental science would make society less idolatrous since it would reveal that the same fundamental structure underlay the law of God and the law of nature. Appreciating the nature of the physical universe as the creation of an all-powerful, governing God would help people to follow the first commandment of worshipping and loving him alone. "This most beautiful system of the sun, planets, and comets," wrote Newton, "could only proceed from the counsel and dominion of an intelligent and Powerful Being."[40] It would also explain the reasons behind the second commandment of understanding our duty "towards one another." Newton's third law of motion conveys the fundamental principle underlying the biblical commandments that deal with the duty between men, according to which action and its consequences were indelibly and accurately balanced. These principles, the third law of motion and the biblical moral one, expressed in a succinct practical idiom a universal theology that governed both the civil and the natural worlds. Newton's scientific work reinforced this theology by offering the means to calculate accurately the outcome of the interactions of physical bodies. His experimental philosophy could thus undertake to liberate men from the shackles of idolatrous ignorance, providing at the same time the practical guidelines necessary for restoring the true worship.

Yet for Newton the similarity between the laws of God and the laws of nature did not end in the concept that the physical universe is the creation of a powerful God and that action and its effects are precisely balanced. His experimental science enabled him to further theorize why God chose to reveal to the Noachides the worship around the Prytanaea and to the Israelite prophets the specific mathematical measurements of the Tabernacle and the Temples. These rites and measurements were designed in accordance with the mathematical formula of the orderly design of the celestial bodies, which Newton was fortunately destined to re-discover in the seventeenth century. All of this was the hidden plan of providence. The differential equations derived from the laws of motion of the designed world alone could not condition the symmetry and harmony found in nature. The fact that symmetrical structures appear everywhere in nature was a sign for Newton of the work and choice of providence. In antiquity God's glory was magnified and celebrated in religious rites that expounded the symmetry and harmony of the order of the design. "Opposite to godliness is atheism in profession and idolatry in action," Newton wrote:

> Can it be by accident that all birds, beasts, and men have their right side and left side alike shaped (except in their bowels); and just two eyes, and no more, on either side of the face; and just two ears on either side [of] the head; and a nose with two holes; and either two forelegs or two wings or two arms on the shoulders, and two legs on the hips, and no more? Whence arises this uniformity in all their outward shapes but from the

[40] Newton, General Scholium, *Principia*, p. 440.

> counsel and contrivance of an Author?... These and suchlike considerations always have and ever will prevail with mankind to believe that there is a Being who made all things and has all things in his power, and is therefore to be feared.[41]

We have seen that for Newton mathematics and symmetry are key factors in God's design of the universe, though they are not the only component governing the design. Like Galileo and Kepler who declared that nature is written in a mathematical language, Newton said in the preface to the *Principia*, "we offer this work as the mathematical principles of philosophy."[42] The book is titled *Principia Mathematica* (not mechanics), because the original design God imprinted upon matter has a mathematical structure of which the solar system is only one expression. At the beginning of Book III, Newton writes, "In the preceding Books I have laid down the principles of philosophy, principles not philosophical, but *mathematical*."[43] The original properties of rays of light also have a mathematical/geometrical design that Newton discovered through the experimental method. In the famous crucial experiment in *Opticks* with the two prisms, Newton held that this experiment was crucial precisely because it revealed the original mathematical design of rays of light. He says, "colors are not *qualifications of light*, derived from refraction or reflections of natural bodies (as it is generally believed), but *original* and *connate properties*, which in divers rays are divers."[44] Similarly in the General Scholium of the *Principia*, he argued that his prior mathematical-physical treatment of motions of bodies revolving around central forces enabled him to show the true original mathematical order of the solar system.[45]

The fact that Newton discovered that material bodies and rays of light have an original mathematical design may not surprise the modern reader, yet the theological implications to be deduced from these assumptions probably will. Let me be more explicit on this point. I am attending to extend studies by many Newtonian scholars,[46] which argue that all aspects of Newton's work were unified inasmuch as they all reflected his religious beliefs. I go further, for I contend that even Newton's mathematics is a part of this unified enterprise and a most essential one since it is intended to expose the hidden design in its most abstract and pure structure. More specifically, I will show that his method of fluxions may be interpreted as a mathematical abstraction of the mechanism responsible for the perseverance and corruption of the original

---

[41] "A Short Scheme of the True Religion," quoted in Brewster, *Memoirs of the Life*, vol. 2, pp. 347-8.

[42] Newton, *Principia*, p. 4.

[43] Newton, *Principia*, p. 319.

[44] Newton, "The New Theory about Light and Colors," *Philosophical Transactions of the Royal Society*, no. 80 (Feb. 19, 1672), pp. 3075-87, p. 3081.

[45] See e.g.: Newton, *Principia*, p. 440.

[46] See discussion on previous chapter.

pristine design, which is sustained constantly by God. There is a problem with such a claim. In the first place, Newton never expressly said this. The closest allusion I could find to the notion that catastrophic events in world history follow mathematical laws appears in his works on the Tabernacle and the Temples (see chapter IX). However, this mathematical lawfulness is not concerned directly with the method of fluxions. While it easy to find citations where Newton says that his natural philosophy is linked to his religious beliefs,[47] he nowhere makes a similar claim about his method of fluxions. He does not say explicitly that his method of fluxions was bound up with his biblical studies, his concept of Providence, or any other religious belief. The most extended discussion of this link is Guiciardini's analysis of the geometrical presentation of the method of fluxions in the *Principia*. Guicciardini devotes a chapter of his book on Newton's *Principia* to the possibility that Newton's geometrical presentations were part of his larger project to recover the ancient wisdom,[48] yet he does not speak of its inner structure as exposing the original design God imprinted upon matter and the mathematical lawfulness of its perseverance and corruption.

Nonetheless, it is well known that in the *Opticks* and the *Principia*, Newton begins with axioms and definitions. The three laws of motion follow a series of eight definitions concerning the quantity of matter and motion, *vis insita*, and a variety of forces. In *Opticks* there is a series of eight definitions on the rays of light, followed by eight axioms. What was the status of these definitions and axioms in Newton's mind? It is accepted that in mathematics axioms and definitions, and the mathematical theorems deduced from them, each have a different status. The theorems are latent in the axioms and definitions. They are not a matter of choice and are not external to the system. Yet the axioms and definitions themselves have a different status, of which Newton was well aware. They are a matter of choice and are external to the mathematical system though they condition everything deduced from them. For example in his mechanics Newton defines the three laws of motion as "the laws and conditions of certain motions, and powers or forces."[49] These principles, according to Amos Funkenstein, show the emergence of a new sense of scientific abstraction, which interprets physical laws as counterfactuals and limiting cases of reality.[50] More to the point, in Newton's mechanics these principles/axioms condition the motion of bodies and the definitions and axioms of Newton's method of fluxions are central to the project of the *Principia*.[51] They are the conditions and principles that describe the rate of

---

[47] For example the General Scholium, on this see: Steve Snobelen, "God of gods."

[48] Niccolo Guicciardini, *Reading the Principia*.

[49] Newton, *Principia*, p. 319, see also discussion on Chapter VI, # iv.

[50] Amos Funkenstein, *Theology and the Scientific Imagination*, (Princeton, 1986), pp. 152-178.

[51] On this see: Richard T.W. Arthur, "Newton's Fluxions and Equably Flowing Time," *Studies in History and Philosophy of Science*, vol. 26 (1995), pp. 323-351; François De Gandt, *Force and Geometry in Newton's **Principia*** (Princeton, 1995); I.B. Cohen, "The *Principia* Universal Gravitation,

fluxions, whether the fluxion is a point, a line, a surface, or a physical body. I cite Newton's own definition from his work on the method of fluxions:

> Mathematical quantities I here consider not as consisting of least possible parts, but as described by a continuous motion. Lines are described and by describing generated not through the apposition of parts but through the continuous motion of points, surface-areas are through the motions of lines... times through continual flux, and the like cases. These geneses take place in the reality of physical Nature and are daily witnessed in the motions of bodies.[52]

Here we can assume that for Newton the axioms and definitions of the method of fluxions function as the principles and conditions of the flux of time and the motion of bodies in space. Thus it is not surprising to find him saying explicitly that:

> geometry and mechanics are joined together in the closest unity, and to speak more truly, they are nothing other than two parts of the one and the same science.... *I say that geometry from its nature is nothing other than the more select part of mechanics, indeed I say that nearly all the mathematical sciences originated from mechanics.*[53]

The geometry Newton employed in the *Principia* was founded, as he pointed out, in a mechanical practice that formed "part of universal mechanics, which accurately proposes and demonstrates the art of measuring."[54] Studying Newton's work as it unfolded in the various drafts and editions of the *Principia,* I.B. Cohen discerned a distinctive "style" structured in three dominant stages of inquiry. Newton first constructed a mathematical set of conditions and explored their mathematical consequences. In the second stage he modified these simplified mathematical conditions, compared them with the relevant phenomena, and introduced additional factors that transformed the original mathematical construct into a physical construct. And finally, in the third stage he "shifts from the level of analogue by applying his analysis and results to the phenomena of the world of observation."[55] The project as a whole

---

and 'Newtonian Style,''' Zev Bechler (ed.), *Contemporary Newtonian Research*, pp. 21-109; D.T. Whiteside (ed.), *The Mathematical Papers of Isaac Newton*, (8 vols'; Cambridge, 1967-81), vol. 8, pp. 92-159.

[52] D.T. Whiteside (ed.), *The Mathematical Papers of Isaac Newton*, vol. 8, p.123.

[53] Newton, *The Mathematical Papers of Isaac Newton*, vol. 7, pp. 339-341. My emphasis.

[54] Newton, *Principia*, on p. 3.

[55] I.B. Cohen, "A Guide to Newton's *Principia*," *Isaac Newton: The Principia*, p. 151.

comprised an attempt to construct an accurate mathematical manual for the analysis of the motion of bodies ruled by specific forces in diverse surroundings. Rather than showing that a given phenomenon formed the conclusion of inferences from mathematical premises, the "Newtonian style" set out to gradually map a wide terrain of phenomena in accordance with Newton's basic laws of motion. Throughout the *Principia*, Newton kept modifying the mathematical and geometrical data from one example to the next, explaining how the universal laws provide the means for accurate representations of different motions in the universe. Disciplined playfulness characterizes Newton's explorations, where each lemma, proposition, and theorem succinctly summarized his attempt at juggling his mathematical apparatus, so to speak, by looking at it from slightly different angles with slightly differing variables.

I am not the first to recognize that the method of fluxions resembles some central features of the laws of motion of bodies and the eight definitions of the *Principia*, yet in a more abstract and general way than in mechanics.[56] However, I go a step further by associating the method with a theological dimension and a speculative argument that Newton held the mathematical components of the material design and the original religion to have a similar structure, so that improper human actions disrupt also the original order of the material design. In addition, I argue that *vis insita*, absolute space and time, and momentary external forces play a role also in the method of fluxions. In the following chapters, I substantiate this claim by contrasting Newton's mathematical method with Leibniz's calculus, discussing the differing metaphysical and physical assumptions that guided Leibniz's mathematics. I point out that Newton's method of fluxions and Leibniz's calculus rely upon different definitions and axioms. The reasons behind choosing these axioms and definitions rather than others are not mathematical but depend upon opposing spiritual intuitions, which I relate to their respective understandings of God's unlimited perspective and the ways open for human beings to share it. Once we are able to recognize the spiritual intuition behind the choice of a certain set of axioms and definitions, we may enter into the heart of the system. In this respect, the axioms and definitions of the method of fluxions and the calculus have an underlying theological structure that captures each thinker's understanding of divinity and matter. For Newton the differential equations of the method of fluxions could not determine the initial conditions of the design. They point out though that God chose to create an ordered symmetry in the design and a "most beautiful system of the sun, planets and comets."[57] Thus God's choice of a certain ordered design is a sign of his dominion, yet nothing conditions him to chose a certain order. For Leibniz, in contrast, the differential equations of the calculus necessarily specify the conditions of the best choice and thus determine and guide the outcome of creation. God is bounded by the

---

[56] De Gandt, *Force and Geometry*; Arthur, "Newton's Fluxions"; Whiteside (ed.), *The Mathematical Papers of Isaac Newton*, vol. 8, pp. 92-159.

[57] Newton, *Principia*, General Scholium, p. 440.

rational logic of the calculus to choose the best. This difference is striking in its theological implications.

For Newton, the created world is a huge designed machine that functions efficiently as long as humans operate it according to the original design given and chosen for them by the governor of the machine. Since human beings are both a part of this design and play a role in preserving it, their actions are of primary importance. When their actions accord with the divine commandments of Noah's simple religion, everything goes smoothly; when their actions ignore or defy the commandments, there are catastrophic repercussions for both the human and material realms. For Leibniz, in contrast, the initial conditions of the created world were chosen according to the most rational laws of nature. As such the actions of human beings are already predetermined according to the best and most rational choice and there is no need for any divine intervention or amending of the design, since evil is but a temporary privation of rational reasoning. The best choice determined in advance that the created world would improve as it progresses.[58]

[58] I am well aware of Leibniz's distinction between absolute necessity and a necessity guided by the best choice – according to the principle of the best. This issue is discussed in length in the following chapters. Human actions are governed by the second necessity. See: Leibniz, *Theodicy: On the Goodness of God the Freedom of Man and the Origin of Evil*, Austin Farrer (ed.), (Illinois, 1990), pp. 59-61.

CHAPTER III

NEWTON'S METHOD OF FLUXIONS

> "I consider time as flowing or increasing by continual flux & other quantities as increasing continually in time & from the fluxion of time I give the name of fluxions to the velocities with which all other quantities increase." (Newton, *The Mathematical Papers of Isaac Newton*, vol. 3, p. 17).

Newton and Leibniz invented the calculus at an early stage of their respective careers, Newton at his home in Woolsthorpe during his *annus mirabilis* (1666, when he was only twenty four years old) and Leibniz toward the end of his stay in Paris (1675, at the age of twenty nine).[1] Both of them developed a sophisticated natural philosophy and became presidents of great scientific academies in their old age. However, their careers followed very different paths. Newton became one of the most influential scientific figures in England, his success and fame reaching international levels, whereas Leibniz even at his greatest moments remained unappreciated by many of his German contemporaries and the wider natural philosophical community. The bitter dispute that arose between them regarding the priority of the calculus in many ways reflects their personal careers. Though Leibniz published his results twenty years before Newton, and developed a more sophisticated algorithm for calculation than Newton did, he was the one accused of plagiarism by the natural philosophical community. The verdict in the dispute was given by a committee of scientists from the Royal Society, of which Newton was president. Leibniz, though president of the Berlin Academy and a foreign member of the Royal Society, was not only accused of plagiarism but abandoned in Hanover by his Duke, afterwards George I, King of England.[2]

---

[1] Leibniz's paper was published in the *Acta Eruditorum*, October, 1684, titled: "Nova Methodus pro Maximis et Minimis." Newton appended his "Tractatus de Quadratura Curvaturum," and "Enumeratio Linearum Tertii Ordinis" in the *Opticks* (1704).

[2] For a more detailed biographical account of the differences between the two men, see: H. G. Alexander (ed.), *The Leibniz-Clarke Correspondence* (New York, 1970), introduction; Ernest Cassirer, "Newton and Leibniz," *Philosophical Review*, vol. 52 (1943), pp. 366-91; Gideon Freudenthal, *Atom and Individual in the Age of Newton: On the Genesis of the Mechanistic World View* (Dordrecht, 1986); Rupert Hall, *Philosophers at War: The Quarrel between Newton and Leibniz* (Cambridge, 1980); J.E. Hoffman, *Leibniz in Paris 1672-76: His Growth to Mathematical Maturity* (Cambridge, 1974); Frank Manuel, *A Portrait of Isaac Newton* (New York, 1968), pp. 321-349; Steven Shapin, "On Gods and Kings: Natural Philosophy and Politics in the Leibniz-Clarke Correspondence," *Isis*, vol. 72 (1981), pp. 187-215; Richard Westfall, *Never at Rest* (Cambridge, 1980), pp. 698-781.

I. PROLOGUE: THE HISTORY OF THE CALCULUS

Many historians of mathematics have written on the calculus, each emphasizing different aspects of its development.[3] "The invention of the calculus, - namely of a method of finding tangents and quadratures which identifies the reciprocity between these two operations –", writes Domenico Bertoloni-Meli, "was the culmination of a process involving several important advances. The establishment of a new, highly abstract, and general form of algebra, the formulation of analytical geometry, and the creation of a variety of techniques for finding maxima, minima, and tangents, [all of these] paved the way to the great inventions by Newton and Leibniz."[4] Indeed, an immense amount of knowledge of the calculus had accumulated before Newton and Leibniz made their syntheses. According to Morris Kline, their calculi were created primarily to treat four types of mathematical problems with which seventeenth century scientists were struggling. The first was, "given the formula for the distance a body covers as a function of time, to find the velocity and acceleration at any instant; and conversely, given the formula describing the acceleration of a body as a function of the time, to find the velocity and the distance traveled." The second type of problem was to find the tangent to a curve. The third was to find the maximum or minimum value of a function, and finally, the fourth was "finding the length of curves" (for example, the distance covered by a planet in a given period of time; the areas; and centers of gravity of bodies).[5] These four problems were considered as distinct before Newton and Leibniz developed their calculi and recognized the generality underlying the separate categories.

Indeed, regarding the category of treating geometrical entities in terms of motion, Newton was the mathematician who made the first significant treatment of general rate problems. He gave a general method for finding the instantaneous rate of change of one variable with respect to another and also showed that the area can be obtained by reversing the process of finding a rate of change. In his mathematical calculations, "motion and time are constantly at work: curves are trajectories traversed by moving bodies; on these curves, points approach each other; the curves themselves are bent, and so forth."[6] Many mathematicians before Newton, such as Galileo Galilei (1564-1642), Marin Mersenne (1588-1648), Gilles Personne de Roberval (1602-1675) and

---

[3] Florian Cajori, *A History of Mathematics* (New York, 1919); C. Boyer, *History of the Calculus* (New York, 1949); M.E. Baron, *The Origins of the Infinitesimal Calculus* (Oxford, 1969); D.T. Whiteside, "Patterns of Mathematical Thought in the Late Seventeenth Century," *Archive for History of Exact Science*, vol. 1 (1961), pp. 179-338; Morris Kline, *Mathematical Thought from Ancient to Modern Times* (New York, 1972).

[4] Domenico Bertoloni-Meli, *Equivalence and priority: Newton versus Leibnitz* (Oxford, 1993), p. 56.

[5] Kline, *Mathematical Thought*, pp. 342-43.

[6] De Gandt, *Force and Geometry*, p. 202.

René Descartes (1596-1650), had already analyzed geometrical curves in motion. However, Newton was the first to analyze figures in motion with respect to finite ultimate ratios.[7] In his calculus, "fluxions express the speed of change of a variable and are finite. They result from variables flowing continuously, almost always with respect to time. Hence kinematics is part of the foundation of the Newtonian calculus."[8]

Regarding the second problem, of finding the tangent to the curve, mathematicians were searching for ways to free the definition of the tangent from physical concepts. Descartes' method was purely algebraic and did not involve the concept of the limit, whereas Pierre Fermat's (1601-1665) method had "the form of the now-standard method of the differential calculus, though it begs entirely the difficult theory of limits."[9] Isaac Barrow's (1633-1677) geometrical methods shortened calculations tremendously by employing the characteristic triangle (Blaise Pascal (1623-1662) had used it even earlier in connection with finding areas), which Leibniz further developed and generalized. Leibniz's main contribution to the definition of tangent was his ability to conceive curves as infinitangular polygons consisting of incomparably many rectilinear segments, while tangents were considered as the prolongation of these segments. Influenced by Pascal's work on harmonic series,[10] Leibniz discovered the existence of a structural resemblance between differences and sums of infinite series and that of tangents and quadratures of geometrical curves. Analyzing the resemblance of the two different realms helped him to develop a coherent theory of differentials and summations. In contrast to Newton's geometrical and kinematic insights, which strictly employ only finite ultimate ratios; Leibniz's thought penetrated deeper into the infinitely small. His calculus did not rely upon geometrical imagination, but upon a mechanical algorithm for calculating differentials and summations.

The problem of the infinitely small grew out of the fourth type of problem, that of finding areas, volumes, and length of curves. The identification of curvilinear areas and volumes with the sum of an infinite number of infinitesimal elements is the essence of Johann Kepler's (1571-1630) method. Galileo also conceived of areas in a manner similar to Kepler. In treating the problem of uniformly accelerated motions, Galileo pointed out that the area under the time-velocity curve is the distance. Bonaventura Cavalieri (1598-1647) further developed Kepler's and Galileo's work into a coherent geometrical method. Cavalieri regarded an area as made up of an indefinite number of equidistant parallel line segments and a volume as

---

[7] Ultimate ratios will be discussed further on in the chapter.

[8] Bertoloni-Meli, *Equivalence and priority*, p.72.

[9] Kline, *Mathematical Thought*, p. 345.

[10] Actually Leibniz was interested in summations and differences already in his dissertation, *Dissertatio de arte combinatoria*. In this dissertation he was trying to work out an alphabet of human thought. See: E.J. Aiton, *A Biography*, (Bristol & Boston, 1985), pp. 18-21, and H.J.M. Bos, "Fundamental Concepts of the Leibnizian Calculus," *Studia Leibnitianna*, Sonderheft 14 (1986).

composed of an indefinite number of parallel plane areas, called respectively indivisibles of area and volume.[11] François De Gandt says that Cavalieri's method was further developed by Evangelista Torricelli (1608-1647), and both methods were well defined and consisted of clearly distinct logical entities. Nonetheless, "from around 1650, the methods of indivisibles were much less firmly characterized; philosophers and scientists thought themselves authorized to speak of these new objects in an extremely vague manner. Cavalieri was referred to, even invoked piously, but seldom read." Further, "the indirect transmission of the theory can explain the great liberty during this era in the usage of indivisibles," by authors as influential as John Wallis (1616-1703), Christiaan Huygens (1629-1695), and Barrow.[12] Indeed both Newton and Leibniz reacted against this ambiguous trend of employing indivisibles uncritically. Newton replaced the vagueness of indivisibles with a rigorous theory founded on ultimate ratios and nascent and evanescent magnitudes,[13] while Leibniz replaced indivisibles with operations of differentiation and integration. Nonetheless, their calculus differs tremendously, both in form and concept.

The relationship among the four distinct problems had been noted and even utilized among leading mathematicians. "For example, Fermat had used the same method for finding tangents as for finding the maximum value of function. Also, the problem of the rate of change of a function with respect to the independent variable and the tangent problem were readily seen to be the same. In fact, Fermat's and Barrow's method of finding tangents is merely the geometrical counterpart of finding the rate of change."[14] Nevertheless, it was not clearly understood that the integral can be found by reversing the differentiation process (namely, finding the anti-derivative). Torricelli came close to grasping this general point, since he saw that in special cases the rate problem was essentially the inverse of the area problem (it was, in fact, involved in Galileo's use of the fact that the area under a velocity-time graph gives distance). "Fermat, too, knew the relationship between area and derivative in special cases but did not appreciate its generality or importance.... In *Geometrical Lectures*, Barrow had the relationship between finding the tangent to a curve and the area problem, but it was in geometrical form, and he himself did not recognize the significance." Finally, "James Gregory (1638-1675), in his *Geometriae* of 1668, proved that the tangent and area problems are inverse problems but his book went unnoticed." In comparison to all previous works mentioned above, Newton's theory of fluxions and Leibniz's calculus clearly demonstrated a "greater generality of method and the

---

[11] Kline, *Mathematical Thought*, pp. 348-350.

[12] De Gandt, *Force and Geometry*, pp. 200-1.

[13] For a clear exposition of the contribution of Newton's strict method to the traditional problems, see: Philip Kitcher, "Fluxions, Limits, and Infinite Littleness: A Study of Newton's Presentation of the Calculus," *Isis*, vol. 64 (1973), pp. 33-49.

[14] Kline, *Mathematical Thought*, pp. 355-356.

recognition of the generality of what has been established in particular problems."[15] Thus Newton's and Leibniz's contribution to the four main problems – rates, tangents, maxima minima, and summation – was reduction and synthesis of all of these problems under the benefice of a new calculus.

II. NEWTON'S METHOD OF FLUXIONS

Newton's method of fluxions is physically oriented and rests upon similarly physical assumptions, which appear in his celebrated *Principia* (1687). As I have shown in the previous chapter, basing myself on I.B. Cohen's work, the theoretical studies in Newton's Principia reveal a highly disciplined craft-work in which mathematical symbols and functions provide the tools for devising increasingly accurate representations of natural processes.[16] The geometry Newton employed in the *Principia* was founded, as he pointed out, in a mechanical practice that formed part of universal mechanics, which accurately proposes and demonstrates the art of measuring. Cohen delineated a distinctive "Newtonian style" structured in three dominant stages of inquiry.[17] The undertaking as a whole comprised an attempt to construct of an accurate manual for the analysis of the motion of bodies ruled by specific forces in diverse surroundings, gradually mapping out a wide terrain of phenomena in accordance with the three laws of motion. In both physics and mathematics, Newton investigated the continuous flow of discrete quantities in relation to a persevering quantity that flows uniformly in relation to the equable flow of time.[18] The intimate relation between the method of fluxions and the motion of physical bodies is expressed in Newton's *De Quadratura Curvaturum*, written around 1704:

> Mathematical quantities I here consider not as consisting of least possible parts, but as described by a continuous motion. Lines are described and by describing generated not through the apposition of parts but through the continuous motion of points, surface-areas are through the motions of lines... times through continual flux, and the like cases. These geneses take place in the reality of physical Nature and are daily witnessed in the motions of bodies.[19]

---

[15] Kline, *Mathematical Thought*, p. 356.

[16] I.B. Cohen, "A Guide to Newton's Principia," *Isaac Newton: The Principia*, pp.148-155.

[17] I.B. Cohen, "A Guide to Newton's Principia," p. 151.

[18] On the development of Newton's method of fluxions, see: Niccollo Guicciardini, *Reading the* ***Principia***, pp. 17-38; regarding the relation between mathematics and physics in Newton's *Principia*, see: idem, pp. 39-98. A precise and clear description on the Newtonian style appears in: Cohen, "A Guide to Newton's *Principia*," *Isaac Newton, The Principia*, pp. 129-138, pp. 148-153.

[19] Isaac Newton, *The Mathematical Papers of Isaac Newton*, vol. 8, p.123 (cited previously). See

On the basis of this text and others, Cohen, De Gandt, and Richard Arthur have argued that Newton's method of fluxions must be understood in relation to the equable flow of absolute time as Newton defines it in the *Principia*.[20] Arthur points out that for Newton duration was not a kind of change in the thing enduring since "the existence of fluents (i.e. flowing quantities) presupposes a temporal flux. A quantity cannot continually increase or diminish in time unless time itself undergoes a continual accretion."[21] In fact, in a draft from 1712, which later appeared anonymously in the *Philosophical Transactions*, Newton wrote:

> I consider time as flowing or increasing by continual flux & other quantities as increasing continually in time & from the fluxion of time I give the name of fluxions to the velocities with which all other quantities increase.... I expose time by any quantity flowing uniformly & represent its fluxion by a unit, and the fluxion of other quantities I represent by any other fit symbols.[22]

If we look at this text carefully, we find that for Newton time is the most fundamental entity and the measure of all quantities. He was consistent on this issue throughout his life. The "continual flux" of mathematical time is in consonance with his definition of the equable flowing duration of absolute time of the *Principia*.[23] Both mathematical and physical times are beyond human reach and can be exposed only by finite quantities, which flow uniformly congruent to the equable flow of absolute time. Thus in mathematics, Newton assigns the fluxion of a chosen uniformly flowing quantity the privileged status of a unit from which all other fluxions are

---

also: "Geometry and mechanics are joined together in the closest unity, and to speak more truly, they are nothing other than two parts of the one and the same science.... I say that geometry from its nature is nothing other than the more select part of mechanics, indeed I say that nearly all the mathematical sciences originated from mechanics." Newton, *The Mathematical Papers of Isaac Newton*, vol. 7, pp. 339-341.

[20] In the next chapter, I present a more detailed discussion of Newton's absolutes. See also: I.B. Cohen, "A Guide to Newton's *Principia*," p. 106, 116; De Gandt, *Force and Geometry*; Richard T. W. Arthur, "Newton's Fluxions," pp. 323-351. The methodological shift that Guicciardini points out in his book does not exclude the concept of an equable flow of a mathematical entity that functions analogous to absolute time. Newton refers to it as part of the synthetic method of fluxions. Guicciardini, *Reading the Principia*, pp.112-115.

[21] Arthur, "Newton's Fluxion", p. 327.

[22] Newton, *The Mathematical Papers of Isaac Newton*, vol. 3, pp. 17-8.

[23] "Absolute, true, and mathematical time," Newton writes in the *Principia*, "of itself, and from its own nature flows equably without regard to anything external, and by another name is called duration." Newton, *Principia*, p.13.

analyzed, whereas in physics, he analyzes accelerated deviations caused by external forces from inertial frames of reference. Furthermore, the homogeneity of all mathematical quantities echoes his understanding of the homogeneity of physical atoms. In both cases the homogeneous investigated quantities (mathematical or material) differ only in their fluxion or trajectories. In other terms, in both mathematics and physics, uniformly flowing relative time is the independent variable from which Newton investigated curves and trajectories.[24] A concise statement of the above is given in his "The Method of Fluxions and Infinite Series":

> We can, however, have no estimate of time except in so far as it is expounded by an equable local motion, and furthermore quantities of the same kind alone, and so also their speeds of increase and decrease, may be compared one with another. For these reasons I shall, in what follows, have no regard to time, formally so considered, but from quantities propounded which are of the same kind shall suppose some one to increase with an equable flow: to this all the others may be referred as though it were time, and so by analogy the name of 'time' may not improperly be conferred upon it.[25]

Once again here and elsewhere, real time is "expounded and measured" by an equably increasing entity. Or in De Gandt's terms: Newton puts "several curves or magnitudes into relation through a very clear system establishing relative location and successive dependence, and among the geometrical magnitudes, there is one that receives a privileged status - it is a displacement that serves as a fundamental variable and which is called time. And as a function of this basic variation, one can compare the variations of other magnitudes, such as occur if the time or the abscissa receives a very small increment."[26] Or in Newton's own terms:

Because we possess no estimation of the time except insofar as it is represented and measured by the intermediary of a uniform local motion, and furthermore because quantities, and their velocities of increase or decrease, can be compared with one another only if they are of the same kind, for this reason I shall not in what follows have any regard for the time taken formally, but among the quantities proposed which are of the same kind I shall suppose one of them to increase according to a uniform fluxion, and I shall relate all the

---

[24] For a fuller exposition of Newton's mathematical time, see: De Gandt, *Force and Geometry*, pp. 159ff.; and Whiteside's illuminating notes on Newton's method of fluxions, in: *The Mathematical Papers of Isaac Newton*, vol. 8, pp. 92-159.

[25] Newton, *The Mathematical Papers of Isaac Newton*, vol. 3, p. 73. This paper was written during 1670-73.

[26] De Gandt, *Force and Geometry*, pp. 199-200.

other quantities to this as if it were time itself, so the name of time can rightly be attributed to it by analogy.[27]

Newton's systematic work, which privileges enduring quantities (quantities generated by an equable flow) over altered ones (quantities whose fluxion keeps changing its rate of flow), is clearly exemplified in the following example taken from *De Quadratura curvaturum.*

Figure 1[28]

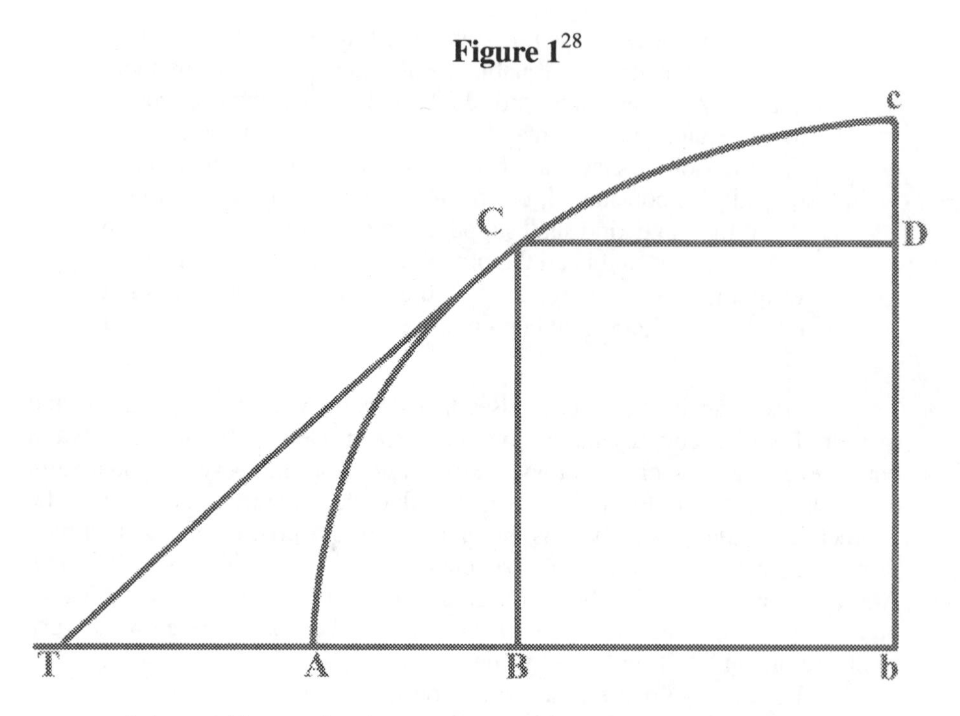

It is manifest at first inspection that these fluxions [i.e. the changing curved-line AC, and the uniform flowing AB], are to one another as the curve's tangent CT, its subtangent TB and its ordinate BC. But were this not evident, the problem would be solved in this manner: let the ordinate BC advance from its place BC to any new place bc and, with the perpendicular CD let fall to bc, the augments of the curve, its abscissa and its ordinate will be Cc, CD and Dc, and the required fluxions will then be to one another in the first ratio of these nascent increments, that is in the ratios of the tangent CT, the subtangent TB and the ordinate BC.[29]

---

[27] Newton, *The Mathematical Papers of Isaac Newton*, vol. 3, p. 72.

[28] Newton, *The Mathematical Papers of Isaac Newton*, vol. 8, p. 98.

[29] Newton, *The Mathematical Papers of Isaac Newton*, vol. 8, p. 95.

Here as elsewhere, most of Newton's procedures preserve close links with geometrical and kinematical intuition.[30] In the above example, Newton constructs a persevering frame of reference from which he can investigate the rate of change of a quantity that does not maintain an equable flow. Time is replaced by a quantity, which flows uniformly - here the fluxion of A along the x-axis. AB becomes the unit of time from which Newton investigates the changing fluxion of the curve AC. Put differently, he situates himself on a uniformly flowing quantity represented by the line AB and while defining the interval AB as the unit (measuring rod) he investigates the rate of change of the curved-line AC. From B he draws an ordinate which cuts the curved line at point C. This move enables him to analyze what is going to happen at the next nascent increment of time in both fluxions - AB and AC.

This step is permissible since Newton considered all "quantities as increasing continually in time." Thus at the same unit of time we have two differing fluxions to look at: AB (the unit) and the curved-line AC; the question investigated is the ratio of these two fluxions to one another at the following nascent interval of time. Everything else that follows is a comparison through geometrical procedures of the fluxion of AB and AC in relation to BC, which becomes bc. Newton's conclusion is that at the first nascent increment of time the relation of AB and the curved line AC to BC is given by the same ratio present in the triangle TBC. That is, the relation between the subtangent TB and the curve's tangent CT to BC is the same as AB and AC is to BC.

Thus without making further assumptions, Newton is able to calculate the momentary change of the fluxion of the curved line AC at point C, avoiding the need to employ the problematic concept of an average motion between the two points composing the curve. Concentrating the analysis on the begotten flux in a nascent moment of time in relation to a quantity representing absolute time helps him to overcome the paradoxes of motion raised by Zeno.[31] The concept of an average motion between points ($\Delta v = \Delta x / \Delta t$) is an artificial human construct that says nothing about the rate of change of the fluxion at a certain specific moment. Newton's interest in calculating the begotten fluent at a specific moment (today called instantaneous velocity) makes it clear that the same procedure needs to be applied for each point on the curved line AC, since the begotten quantity (instantaneous velocity) at each point is analyzed in relation to an external inertial frame of reference representing time. The relation investigated is not that between the separate points that compose the curve, but the momentary deviation of the curve at point C in relation to an

---

[30] Newton was very insisted regarding his geometrical approach and avoidance of indivisibles. For example, he says in *De Quadratura*: "In finite quantities, however, to institute analysis in this way and to investigate the first or last ratio of nascent or vanishing finites is in harmony with the geometry of the ancients, and I wanted to show that in the method of fluxions there should be no need to introduce infinitely small figures into geometry." Newton, *The Mathematical Papers of Isaac Newton*, vol. 8, p. 129. On Newton's use of the geometry of the ancients, see: Guicciardini, *Reading the Principia.*

[31] On Zeno's paradoxes of motion and Newton's elegant solution, see: Yakir Shoshani, *Thoughts on Reality* (Hebrew), (Tel-Aviv, 1999) pp. 62-70.

external uniformly flowing quantity (AB) representing time.

In his mathematics and physics, the frame of reference from which Newton calculates the begotten fluents is almost always an equably flowing quantity representing time. Quantities flow within time's equable duration as long as they are not forced (momentarily) to change their rate of fluxion. Furthermore, as will be discussed below, in physics, Newton considers a deviation from the equable flow of any quantity to be secondary in most cases to the persevering inertial state, and to be caused by momentary external interactions. A curved fluxion is never generated in Newton's physics due to an internal built-in lawfulness that can be understood by itself without relation to an external inertial frame of reference. The cause of deviation is an external and momentary disruption (that he calls force) and must be calculated over and over again in relation to an external frame of reference (called time).[32] Even central forces, such as gravity, which are continuous, are conceptualized and analyzed as momentary forces.[33]

Newton's systematic work, privileging enduring quantities (quantities generated by an equable flow) over altered ones (quantities whose fluxion does not persevere in the same state) is again exemplified by the following example taken from his algebraic investigations of fluxions. Later on in *De Quadrata Curvaturum*, he calculates the rate of change of two algebraic quantities, x and x^n, and shows that the rate of change of the latter to the first is nx^{n-1} to 1. Once again, in this algebraic problem, Newton first identifies a persevering quantity represented by x. Then, from the equable flow of x he calculates the rate of change which a non-uniform quantity x^n undergoes at the first nascent increment of time:

> Let the quantity x flow uniformly and the fluxion of the quantity x^n need to be found. In the time that the quantity x comes in its flux to be $x+0$, the quantity x^n will come to be $(x+0)^n$, that is (when expanded) by the method of infinite series:
>
> $$x^n + n0x^{n-1} + 1/2(n^2 - n)0^2x^{n-2} + ...;$$
>
> and so the augments 0 and $n0x^{n-1} + 1/2(n^2 - n)0^2x^{n-2} + ...$ are one to the other as 1 to $nx^{n-1} + 1/2(n^2 - n)0x^{n-2}$... Now let those augments come to vanish and their last ratio will be 1 to nx^{n-1};

[32] For an exposition of Newton's geometrical procedures for investigating fluxions, see: De Gandt, *Force and Geometry*, pp. 209-241. For a further discussion of Newton's way of calculating curvatures which are not in the form of a second order differential equation, see: J. Bruce Brackenridge, "The Critical Role of Curvature in Newton's Developing Dynamics," and A. Rupert Hall, "Newton and the Absolutes: Sources," both papers in: P.M. Harman and Alan E. Shapiro (eds.), *The Investigation of Difficult Things: Essays on Newton and the Exact Sciences in Honour of D.T. Whiteside* (Cambridge, 1992), pp. 231-61.

[33] On this subject, see: I.B. Cohen, "Newton's Second Law and the Concept of Force in the *Principia*," Robert Palter (ed.), *The Annus Mirabilis of Sir Isaac Newton* (Cambridge, 1970), pp. 186-91.

> consequently the fluxion of the quantity x is to the fluxion of the quantity x^n as 1 to nx^{n-1}.[34]

Put differently, Newton constructs here a persevering frame of reference (x) from which he can investigate the rate of change of a quantity, which does not persevere in its equable flow (x^n). Following, he compares their augments when vanishing to their last ratio and concludes that "the fluxion of the quantity x is to the fluxion of the quantity x^n as 1 to nx^{n-1}" or in more general terms:

> And hence in equations involving but two unknowns, one of which is a uniformly flowing quantity and the second is any fluxion you please of another fluent quantity, the second fluent can be found through the quadrature of curves.[35]

Thus we see that even in his algebraic calculations, Newton investigates the rate of change of fluxions in relation to an external inertial frame of reference. Always privileging quantities generated by an equable flow (in this case x) over quantities whose fluxion does not persevere in the same state (in this case x^n). This idea is again expressed in *De Quadratura*, where he defines fluxions as "the speeds of motion or increment by which [quantities] are generated," whereas fluents are "the so born" quantities. Thus "fluxions are very closely near as the augments of their fluents begotten in the very smallest equal particles of time: so to speak accurately, indeed, they are in the first ratio of the nascent augments, but they can, however, be expressed by any lines whatever which are proportional to them."[36]

Why is it so important for Newton to discover what happens to the rate of a fluxion at the "smallest equal particle of time?" Why does he concentrate his efforts on deciphering an event occurring in such a nascent or evanescent time? It seems to me that we can get an insight into his mathematical intuition if we interpret his motives as the search for an explanation of the hidden mechanism of the worldly design of bodies. The operation of forces on bodies is not something that we observe in nature.[37] Forces are constructs of the human mind to explain the reason for an observed change seen or experienced in the world. The observed change gives us information regarding a hidden mechanism inherent in bodies that resists the observed change. How can the mechanist decipher and expose this mechanism? All he observes is a continuous displacement of the moving body

[34] Newton, *De Quadratura*, (1704), *The Mathematical Papers of Isaac Newton*, vol. 8, pp. 127-29.

[35] Newton, *De Quadratura*, (1704), *The Mathematical Papers of Isaac Newton*, vol. 8, p. 155.

[36] Newton, *The Mathematical Papers of Isaac Newton*, vol. 8, p.123-5.

[37] I owe the following discussion on forces to Yakir Shoshani.

from point A to point C, but from that observation it is not clear how he will be able to calculate the rate of change the quantity undergoes and its resistance to the change.

For this an analysis of proportions between quantities, representing an observed effect, is essential. Neither God nor the inner hidden mechanism of bodies is present or visible to the eyes of humanity, yet God's works and operations are open to human experience. Forces, likewise, are not present to the human gaze they are spiritual constructs, yet their operations and effects are experienced and can be calculated within an artificial set of experimentation that has control over the investigated parameters. This is precisely what Newton's famous second law of motion calculates. The law ($F \propto a$) shows that forces are always proportional to acceleration and have the same direction. Thus even though the force is a hidden cause the effect (the acceleration, that is, the change of the rate of motion of a body) is, nonetheless, observed and the above proportion can be calculated. From this general observed proportion of force and acceleration, Newton defines both a general law of motion governing bodies and a hidden material quality that due to the observed proportion can be quantified and called mass. Thus $F=ma$. Such a constant can always be defined when we have two proportional magnitudes, $A \propto B$, in this case $A/B=K$, where K is a constant. In order to calculate the implications of such a general law of nature Newton studies in the *Principia* the displacement of this newly defined quantity (mass) under different conditions. He calculates over and over again the observed rate of change of motion that a body undergoes in any nascent moment due to the operation of differing external forces upon different masses in relation to absolute space and time. The absolute frames of reference of space and time enable him to calculate the momentary change. This way of experimentation makes it possible for him to discover the inner mechanism of bodies, although inertia and forces are never observed directly in nature.

This is what he does when he first introduces in the *Principia* the mathematical operations required for such calculations of momentary changes in the fluxion of any given quantity. Newton published his theory on the first ratio of nascent augments eighteen years before the text, which I analyzed above. The first presentation of ultimate ratios appeared in the *Principia* at the beginning of Book I, immediately after the definitions and axioms of the physical entities. Here quantities vanish into a curve from above and below the curve.[38] I believe that these quantities also represented for him the reaction of any inertial mass to a change of the body's passive state due to an outside force exerted upon it momentarily, yet continuously.

In modern terms we can say that Newton here studied the displacement of the curve x as a function of the time t emerging from the forces acting on the moving body, so that he could observe the response of the body to its momentary external interactions. He needed to know $\Delta x/\Delta t$ when the time $t \rightarrow 0$ (what he called ultimate ratios though he did not yet have the modern term of a limit of a function) because only at that evanescent or

---

[38] You can find the illustration as figure 2 later on in the chapter.

nascent moment when a parallelogram merges into the curve from below or above could he calculate the precise rate of change of the motion the quantity undergoes in relation to the absolutes. There was nothing new in his geometrical illustration.[39] Indeed it resembled the work of other contemporary mathematicians, which modified the Greek method of exhaustion.[40] Nonetheless, Newton gave the approaching parallelograms the unique role of first and last ratios of evanescent and nascent quantities. The ultimate ratio functions as the limit for two approaching parallelograms, one below the curve, the other above. The parallelogram above the curve represents a nascent quantity, the figure beneath it an evanescent quantity. When additional instantaneous evanescent and nascent parallelograms are produced the ultimate ratio of all parallelograms represents a curve, which limits two respective finite polygons. What the method of fluxions investigates is the ultimate ratio at which each evanescent quantity vanishes into the curve, and from which each nascent quantity is generated. Only at each of these vanishing moments is the studied motion no longer an average motion between two points on the curve but the precise begotten fluent at that nascent moment of time in relation to an absolute frame of reference. If space and time did not represent absolute frames of reference it would have not been possible for Newton to investigate what happens to the motion at that specific vanishing moment.

In what follows I would like to point out an analogy between the moment before nascent and evanescent quantities vanish into ultimate ratios and the dual reaction of the persevering and protective nature of the passive force - *vis insita* – of bodies to any change of their inertial state. This analogy is of primary importance for the argument of the book since without considering Newton's strong preference for quantities that flow equably in relation to absolute time over those whose fluxion undergoes change, I will not be able to argue that the inherent hidden mechanical design of bodies had for him a structure that gave content and explanation to the true worship of the original religion of Noah and his sons around the sacrificial fire.[41]

Newtonian physics assumes the existence of two separate ontological entities - absolute space (and time) and discrete "permanent particles," called atoms:

> All these things consider'd, it seems probable to me, that God in the Beginning form'd Matter in solid, massy, hard, impenetrable, moveable Particles, of such Sizes and Figures, and with such other Properties, and in such Proportion to Space, as most conduced to the End for which he form'd them; and that these Primitive Particles being Solids, are

[39] The illustration appears as figure 2, below.

[40] See e.g., Kline, *Mathematical Thought*, pp. 351-2.

[41] For a historical and different interpretation on first and ultimate ratios and *vis insita*, see: Cohen, "A Guide to Newton's *Principia*," *Isaac Newton: The Principia*, pp. 96-101, pp. 129-131.

> incomparably harder that any porous Bodies compounded of them; even so hard, as never to weak or break in pieces; no ordinary Power being able to divide what God himself made one in the first creation.[42]

Newtonian atoms, in themselves, are completely separated entities with rigid boundaries that enable them to remain separated from anything other than themselves. Atoms do not contain any information or qualities regarding their inter-relatedness with anything external to themselves. Their non-elastic composure suits well with the category of discrete quantities continuously flowing in space.[43] Bodies are a composition of these atoms therefore they can also be treated as quantities and treated mathematically. Newton's first definition in the *Principia* is precisely on the quantity of this quality he calls matter. He defines this new category in order to give mathematical content to his newly defined concept called mass.[44] The most interesting feature of the concept of mass is its resistance to any change of its physical state of rest or "moving uniformly forward in a right line" and the fact that its reactions to a change of its state are governed by mathematical lawfulness. As long as no external forces disrupt a massive body (composed of atoms) the body perseveres in its inertial state thanks to its internal passive force called *vis insita*. This passive force assures that "every body perseveres in its state of rest or uniform motion in a right line, unless it is compelled to change that state by forces impressed thereon."[45]

This innate passive force (which is a crucial component of the original design of matter) is in charge of resisting any intrusion or interruption to the body's momentary physical state. Like forces, the *vis insita* of bodies is never observed in nature yet we know of its existence since it is exerted when "another force, impressed upon it, endeavors to change its condition; and the exercise of this force may be considered as both impulse and resistance."[46] The exertion of the *vis insita* of a body is always *two-fold* -- impulse and resistance. Impulse is an external combative reaction to the intruder, resistance is an internal defensive and protective response. Crudely put, if Newton's atoms could have contrived never to interact with other entities they would have continued to flow equably with God's eternal duration. However, in the Newtonian world the atom's equable flow is constantly being interfered with

---

[42] Isaac Newton, *Opticks*, p. 400.

[43] See the above cited text: "God in the beginning *form'd* Matter in solid, massy, hard, impenetrable, moveable Particles," *Opticks*, p. 400.

[44] On the particularity and newness of Newton's notion of mass, see: Cohen, *Introduction to Newton's Principia*.

[45] Newton, *Principia*, Book I, Law I.

[46] Newton, *Principia*, definition III.

and deflected due to material interactions.[47]

Returning to ultimate ratios, Newton tells us in the *Principia* that "quantities and the ratios of quantities, which in any finite time converge continually to equality, and before the end of that time approach nearer to each other than by any given difference, become ultimately equal."[48] He then presents a diagram of a curve approached by parallelograms beneath and above the curve:

Figure 2

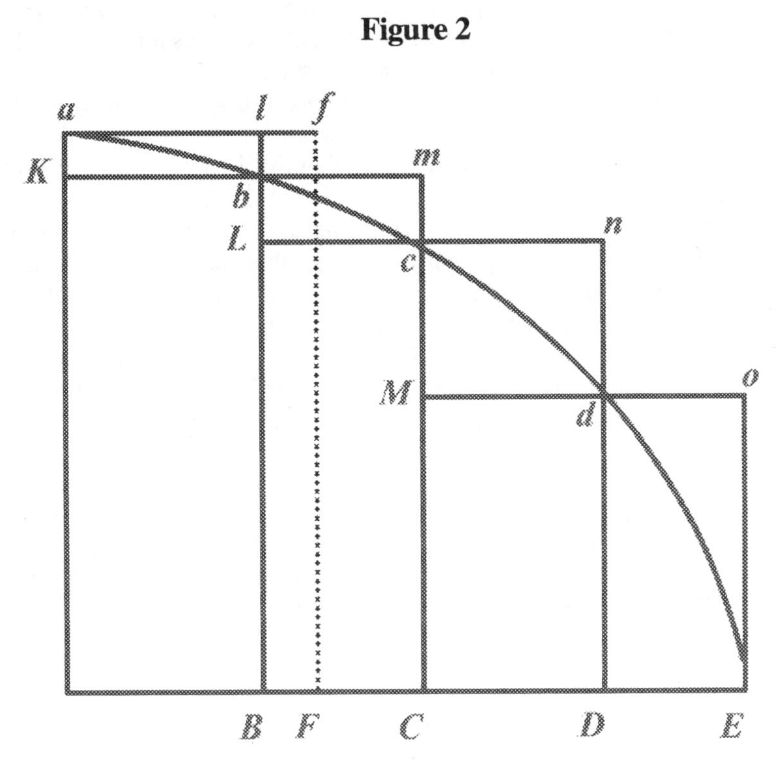

Did not Newton base this Lemma upon the same internal aptitude of atoms to persevere in their private state and strike back if interrupted by other bodies through momentary bouts of impulse and resistance due to their *vis insita*? From the point of view of the body intruded upon, the aggressive external force impels the body's own boundary from outside inward, whereas the body's defensive response is directed from inside out, to avoid any yielding of its boundaries. Indeed the qualitative resemblance between evanescent and

---

[47] See e.g. Newton, *Opticks*, p. 401: "[T]hese particles have not only *vis inertiae*, accompanied with such passive Laws of Motion as naturally result from that force, but also that they are moved by certain active Principles, such as that of Gravity and the Cohesion of Bodies [the cause of interaction]."

[48] Newton, *Principia*, Book I, Section I, Lemma I.

nascent qualities and the dual reaction of *vis insita* is worth further consideration. The method of fluxions investigates the "ratio of quantities [rectilinear figures which are represented by each separate parallelogram] not before they vanish, nor afterwards, but with which they vanish,"[49] while in physics one calculates the momentary alteration of the body's persevering state as a reaction of *vis insita* to an instantaneous external force. In mathematics one investigates a series of momentary evanescent and nascent quantities merging into a curvilinear limit,[50] whereas in physics one searches for the dual reaction of *vis insita* toward any instantaneous impelling force. To be more precise: evanescent quantities are annihilated when reaching the limit/curve from beneath, as the body's persevering state is disrupted from its inertial path by an intruding impelling force; while nascent quantities are generated upward analogously to the reaction of the resistance of *vis insita* toward the instantaneous intruder and the immediate acquiring of a new inertial state.[51] The fact that Newton employs two parallelograms, one beneath and one above the curve, though only one set of parallelograms is needed for calculating the ultimate ratio supports my argument that this diagram has a structure analogous with the reaction of *vis insita* to a continuous intruding force. Another support for the argument is the fact that most mathematicians of his time represented infinitesimal quantities with only one set of parallelograms.

Why then does Newton give such a privileged status to a quantity, which flows equably like absolute time, and not to a fluxion, which continuously alters its persevering duration? Before discussing this issue, we shall analyze Leibniz's calculus in order to further appreciate Newton's non-trivial solution.

---

[49] Newton, *Principia*, p. 39.

[50] "And therefore these ultimate figures [i.e. the curve limiting the polygons] are not rectilinear, but curvilinear limits of rectilinear figures." Newton, *Principia*, Book I, Lemma III.

[51] An even clearer formulation of the dual nature of *vis insita* can be found in Newton's earlier notes. There the dual reaction of *vis insita* is not presented as a passive reaction but as an active one: "Force is the casual principle of motion and rest and is either something external which, impressed in a certain body, either generates or destroys its motion, or at least to some extent changes it; or it is the internal principle by which the motion or rest imprinted on the body is conserved, by which every entity endeavored to persevere in its actual state, and oppose itself to any impediment." MS.ADD.4003, def. 5, John Herivel (ed.), *The Background to Newton's Principia* (Oxford, 1965), p. 231.

CHAPTER IV

LEIBNIZ'S CALCULUS

> "[T]he rules of the finite are found to succeed in the infinite.... And conversely the rules of the infinite apply to the finite... This is because *everything is governed by reason*; otherwise there could be no science and no rule, and this would not at all conform with the nature of the sovereign principle." (Leibniz, "Letter to Varignon," 1702, *Philosophical Papers and Letters*, p. 544)

Coming now to Leibniz, we encounter an entirely different calculus, associated with opposing notions of time and continuity. In contrast to Newton's fluxions and ultimate ratios, Leibniz's calculus deals with series and differentials.[1] A Leibnizian differential (dx) is a difference between two neighboring variables on a continuously increasing quantity. For Leibniz,

> *dx* means the element, that is the (instantaneous) increment or decrement, of the (continually) increasing quantity x. It [*dx*] is also called a difference between two proximate *x's* which differ by an element (or by an unassignable), the one originating from the other, as the other increases or decreases (momentarily).[2]

For a better understanding of Leibniz's intuitions we need to decipher the meaning he assigned to his differentials. On these issues he wrote to John Wallis: "[T]he consideration of differences and sums in number sequences has given me my first insight, when I realized that differences correspond to tangents and sums to quadratures."[3] In his essay "The History and Origin of

---

[1] For greater detail than treated here regarding the comparison, see: Bertoloni-Meli, *Equivalence and priority*; De Gandt, *Force and geometry*, pp. 215-17, 249; Guicciardini, *Reading Newton's* **Principia**, pp. 136-168.

[2] Leibniz, C.I. Gerhardt (ed.), *G.W Leibnitz: Mathematische Schriften*, (7 vol's; Hildeshein, 1854 reprinted 1975), vol. 7, pp. 222-3. It is important to note that this definition of *dx*, as a first-order differential, involves a fundamental indeterminacy, since the choice of the difference is always arbitrary. On this issue see H.J.M. Bos, "Differentials, Higher-Order Differentials and the Derivative in the Leibnizian Calculus," *Archive for History of Exact Sciences*, vol. 14 (1974), pp. 1-90.

[3] Leibniz to Wallis, 28 May, 1697, *Leibnitz: Mathematische Schriften*, vol. 4, p. 25, translated by H.J.M. Bos, "Fundamental Concepts of the Leibnizian Calculus," *Studia Leibnitianna* Sonderheft 14,

the Differential Calculus," written in 1714, as an account of the development of his own thinking,[4] Leibniz informs us that he became particularly interested in the properties of numerical sequences and the sums or differences of the terms in the sequence early on in his Dissertation. There he noticed that the operations of addition and subtraction produce two new sequences, one of sums and the other of differences, and that both new sequences are reciprocally related, yielding the original sequence. Continuing to explore this theory in the 1670's while applying it to the study of curves, Leibniz noticed that if he considered the differences between the terms of the sequences of ordinates and abscissa as negligible with respect to finite quantities, yet unequal to zero, he could make extrapolations from the calculus of these finite sequences to the actually infinite.[5] Thus he concluded that finite approximations give way to exact results when the curve is treated as an infinitangular polygon:

> I feel that this method and others in use up till now can all be deduced from a general principle which I use in measuring curvilinear figures, *that a curvilinear figure must be considered to be the same as a polygon with infinitely many sides.*[6]

Leibniz maintained that this infinitangular polygon coincides with the curve; and that its infinitely small sides if prolonged form tangent lines to the curve. More precisely, as Douglas Jesseph argues, Leibniz thought that:

> [I]n the infinite case, the differences *dx* and *dy* represent infinitesimal increments in the abscissa and ordinate of the curve, and the area beneath the curve can be interpreted as composed of infinitely narrow parallelograms of the form *ydx*. The infinite sums of such parallelograms will then give the true area, while the ratio between *dx* and *dy* gives the slope of the tangent. [Nonetheless, the differences *dx* and dy are *variable* quantities, and] they can be thought of as ranging over infinite sequences of values of *x* and *y* which are infinitely close to one another. Depending upon the nature of the curve, the infinitesimal quantities *dy* and *dx* can stand in any number of

---

1986.

[4] J.M. Child (ed.), *The Early Mathematical Manuscripts of Leibniz* (Chicago, 1920), pp. 22-58. This piece should be read carefully since it was written at the height of the priority dispute with Newton.

[5] On the issue of extrapolation to the actually infinite, see: H.J.M. Bos, "Differentials", pp. 1-90, especially, pp. 13-16.

[6] Leibniz, "Additio ... de Dimensionibus Curvilineorum," *Acta Eruditorum* (December, 1684), pp. 585-87, *Leibnitz: Mathematische Schriften*, vol. 5, p. 126, translated by H.J.M. Bos, "Differentials", p. 14.

> different relations, and because these quantities are themselves variables, it makes sense to inquire into the rates at which they vary. The second-order differences *ddx* and *ddy* can be introduced as infinitesimal differences between values of the variables *dx* and *dy*, and similar considerations would allow the construction of a sequence of differences of ever-higher orders.[7]

To present Leibniz's position in more simplistic terms, let us consider the following basic sequence of squares:

0, 1, 4, 9, 16, 25, 36

The first differences are

3, 5, 7, 9, 11

The second differences are

2, 2, 2, 2, 2

The third differences are

0, 0, 0

Leibniz himself noted the vanishing of the second differences for the sequence of natural numbers, the third differences of squares, and so on, as early as 1666, in his doctoral thesis, *De Arte Combinatoria* (*On the Art of Combinations*).[8] He also observed that if the original sequence starts from 0, the sum of the first differences is the last term of the sequence. He extrapolated from this finite sequence to the actually infinite, observing that the foundation of the infinitesimal calculus is the principle that "differences and sums are the inverse to one another, that is to say, the sum of differences of a series [sequence] is a term of the sequence, and the difference of sums of a sequence is itself a term of the sequence; and I enunciate the former thus, $\int dx=x$, and the latter $d\int x=x$."[9] Leibniz's interest in the fact that differences and sums are reciprocal to one another enabled him, in his first publication in the *Acta Eroditorum* paper of 1684, to offer rules for these new infinitely small mathematical variables.[10]

Nonetheless, in this paper he still needed to clarify what he meant by the differential, especially after claiming that the tangent is the prolongation of the infinitesimal side of a polygon equivalent to the curve.[11] As H.J.M. Bos has

[7] Douglas M. Jesseph, "Leibniz on the Foundation of the Calculus: The Question of the Reality of Infinitesimal Magnitudes," *Perspective on Science*, vol. 6 (1998), pp. 6-40.

[8] C.I. Gerhardt (ed.), *Die Philosophischen Schriten von G.W. Leibnitz*, (7 vol's; Hildeshein, 1849-63 reprinted 1971), vol. 4, pp. 27-102.

[9] Leibniz, "Elements of the New Calculus for Differences and Sums," *The Early Mathematical Manuscripts of Leibniz.*

[10] "Nova Methodus pro Maximis et Minimis," *Acta Eroditorum*, vol. 3, (1684), pp 467-73, *Leibnitz: Mathematische Schriften*, vol. 5, pp. 220-26.

[11] "Nova Methodus pro Maximis et Minimis," *Leibnitz: Mathematische Schriften*, vol. 5, p. 223.

shown, Leibniz, like other seventeenth century mathematicians, treated curves as if they embodied relations between several variable geometrical quantities defined with respect to a variable point. Curves did not represent a graph of a modern function [x➔y(x)], where x is the independent variable. Indeed, Leibniz's differentiation and summation do not act on functions of an independent variable, but on variable geometrical quantities, such as "ordinate, abscissa, arclength, radius, polar arc, subtangent, normal, tangent, areas between curve and axes," and other variables. The differential itself is a variable and differentiation is an operation assigning variables to variables. Thus, says Bos, "[d]ifferentials have different values according to where in the geometrical figure they occur; although infinitely small they have the same characteristics which make ordinate, abscissa etc. variables." Further, differentials hardly ever occur as single entities. Instead, "they are ranged in sequences along the axes, the curve and the domains of other variables; they are variables, themselves depending on other variables involved in the problem, and this dependence is studied in terms of differential equations." [12]

Furthermore, "Leibnizian differentiation," writes Bertoloni-Meli, "associates to a variable x another variable dx infinitely, or better incomparably, small with respect to it, and conversely to integration. Differentiation and integration are operations on variables and change the order of infinity, not the dimension of the variable. The differential of a length is an incomparably small length, the integral of an incomparably small velocity is a finite velocity."[13] Thus, it is important to note that this definition of dx, as a first-order differential, involves a fundamental indeterminacy, since the choice of the difference is always arbitrary. Differentials are indeterminate, since it is possible to choose the dx and the dy to be constant or variable according to the rule with which the sides of the infinitangular polygon have been chosen. Thus "differentials are indeterminate because the sequences of the relevant variable or the associated polygon can be chosen in infinitely many ways. Moreover, differentials can be given an arbitrarily small values in the calculations so that by neglecting them – in the appropriate circumstance– the error in the result is less than any given quantity."[14]

Tangents are a specific class of Leibniz's differentials. On this issue he wrote:

> [T]o find a tangent is to draw a straight line which joins two points of the curve which have an infinitely small distance, that

---

[12] Bos, "Differentials," pp. 5, 17.

[13] Bertoloni-Meli, *Equivalence and Priority*, pp. 66, 68.

[14] Bertoloni-Meli, *Equivalence and Priority*, p. 68. See also Bos's exposition on this issue, in "Differentials." In Leibniz's terms, "[I]f someone does not want to employ *infinitely small* quantities, he can take them to be as small as he judges sufficient to be incomparable, so that they produce an error of no importance and even smaller than any given [error]." "Treatise on the Causes of Celestial Motions," *Leibnitz: Mathematische Schriften*, vol. 4, p. 150.

is the prolonged side of the infinitangular polygon which for us is the same as the curve.[15]

Figure 3

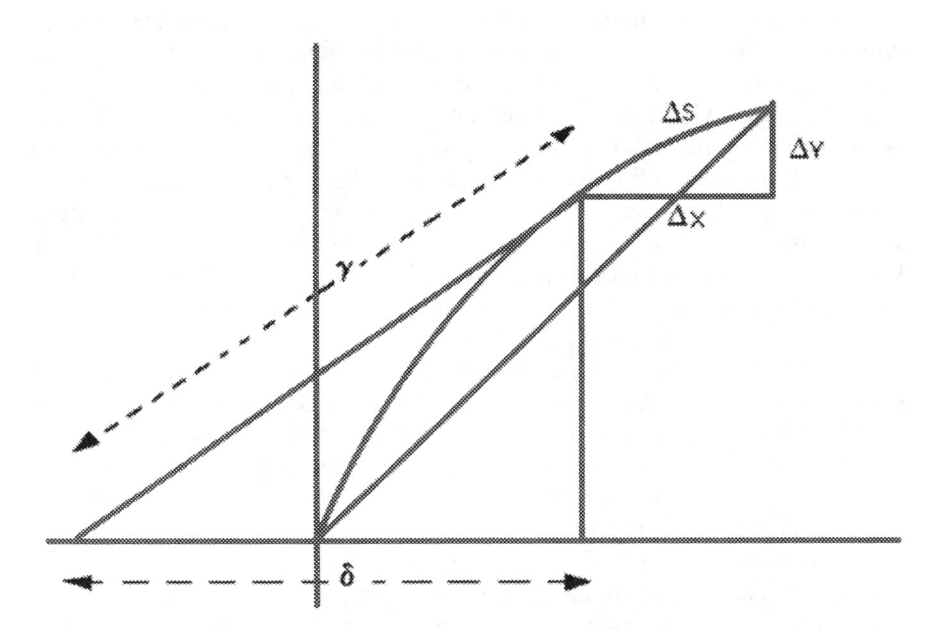

This definition introduces the characteristic triangle (of Pascal), to which Leibniz gave a differential meaning. In the finite display (see the figure), the ratios Δx:Δy:Δs are approximately equal to the ratios σ:y:τ of subtangent, ordinate and tangent. Nonetheless, in the Leibnizian extrapolation to the actually infinite the triangle becomes a differential triangle with sides *dx, dy, ds.* The hypotenuse of the differential triangle is a side of the infinitangular polygon, and therefore, if prolonged, it forms a tangent line to the curve.[16]

Let us look more carefully at the significance of the extrapolation from Δs to *d*s. My understanding is that Leibniz grasped the tangent as an expression of a hidden infinitesimal divine regularity, which generates the most determined path from one infinitesimally close point A to its successive point B (i.e. an Euclidean straight line, represented by the convergence of the differential triangle with the curve).[17] On these issues he wrote to Johann

[15] Leibniz, "Nova Methodus pro Maximis et Minimis," *Leibnitz: Mathematische Schriften,* vol. 5, p. 223.

[16] The figure and explanation are taken from Bos, "Differentials", p.18.

[17] "Hence it is very clearly understood that out of the infinite combinations and series of possible things, one exists through which the greatest amount of essence or possibility is brought into

Bernoulli (1630-1677): "[t]he real infinite, is perhaps the absolute itself, which is not made of parts but which includes beings having parts, in an eminent way and in proportion to the degree of their perfection."[18] By perfection, Leibniz meant "nothing but quantity of positive reality taken strictly, when we put aside the limits or bounds in the things, which are limited. But where there are no bounds, that is, in God, perfection is absolutely infinite."[19] Perfection is a divine attribute, yet human beings' understanding is finite and limited compared to God's infinite perfection. A good example of this difference is the following: Humans, writes Leibniz, "do not have any *idea* of a circle, such as there is in God, who thinks all things at the same time." Nonetheless, rational beings can grasp perfection in a limited way since "there is in us an image of a circle, and also the definition of a circle, and there are in us ideas of those things, which are necessary for a circle to be thought. We think about a circle, we provide demonstrations about a circle, we recognize a circle: its essence is known to us – *but only part by part.*" It is only "if we were to think the whole essence of a circle at the same time, then we would have the idea of a circle. Only God has the idea of composite things."[20]

Thus from human beings' limited perspective the geometrical curve (like the circle in the above example) seems composed of a finite sequence. It is only with the assistance of extrapolation of the finite sequences to the actually infinite that mathematicians gain a momentary divine unlimited perfect perspective and discover an infinitangular polygon residing within the curve. Therefore, it seems more then plausible to me, that Leibniz understood his calculus to be a divinely useful discovery that assists human beings in gaining a wider perspective on the laws governing series. After all he believed that "everything [that] follows the principle of perfection is infinitely more useful to man, and even to the state, than all that serves the arts. Discoveries useful to life, moreover, are very often merely the corollaries of more important insights; it is true here too that those who seek the kingdom of God find the rest on their own way."[21]

Indeed, the most important insight of the calculus is the realization

---

existence.... [Thus] assuming there is to be motion from one point to another without anything more determining the route, that path will be chosen which is easiest or shortest." Leibniz, "On the Radical Origination of Things," 1697, *Die Philosophischen Schriften von G.W. Leibniz,* vol. 7, pp. 302-8 translated in: *G.W. Leibniz: Philosophical Papers and Letters,* p. 487.

[18] Leibniz to Johann Bernoulli, 7 June, 1698, in *G.W Leibniz: Mathematische Schriften*, vol. 3, p. 500, translated by Miklos Horvath, "On the Attempts Made by Leibnitz to Justify his Calculus", *Studia Leibnitia.* Band XVIII/1, (1986), pp. 60-71, p.64.

[19] Leibniz, "Monadology," & 41, 1714, in *Die Philosophischen Schriften von G.W. Leibniz,* vol. 4, pp. 607-23, translated in: *G.W. Leibniz: Philosophical Papers and Letters*, pp. 646-7.

[20] Leibniz, "On Mind, the Universe, and God," 1675, Daniel Garber and Robert C. Sleigh (eds.), *De Summa Rerum: Metaphysical Papers (The Yale Leibniz)* (New Haven, 1992), p. 5. My emphasis.

[21] Leibniz, "Tentamen Anagogicum," 1696, *Die Philosophischen Schriften von G.W. Leibniz,* vol. 7, pp. 270-1 and *G.W. Leibniz: Philosophical Papers and Letters*, p. 477.

that "infinities and infinitesimals are grounded in such a way that everything in geometry, and even in Nature, takes place as if they were perfect realities."[22] Differentiation is a rational operation assisting the mathematician to discover the hidden infinitesimal law of the generation of curves. "[T]he rules of the finite," writes Leibniz, "are found to succeed in the infinite.... And conversely the rules of the infinite apply to the finite... This is because *everything is governed by reason*; otherwise there could be no science and no rule, and this would not at all conform with the nature of the sovereign principle."[23] Put differently, Leibniz understood the curve to be a confused finite expression of a hidden and most determined infinitesimal regularity. This concealed infinitesimal regularity was unfolded through the operation of differentiation. For, as Leibniz said clearly in his celebrated "Principles of Nature and of Grace, Based on Reason":

> One could learn the beauty of the universe in each soul if one *could unravel all that is rolled up in it* but that develops perceptibly only with time. But since each *distinct perception* of the soul includes an *infinity of confused perceptions* which envelop the entire universe, the soul itself does not know the things which it perceives until it has perceptions which are distinct and heightened.[24]

Thus I maintain Leibniz believed that when curves are perceived more distinctly (not confusedly) through operations of differentiation they are actually perceived as infinitangular polygons and not as finite curves. The infinitesimal paths from one point to the other are, so to speak, "rolled up in" the finite sequence of the original curve developing analogously to the way the infinite unnoticed (confused) perceptions "envelop the entire universe." Time, in Leibniz's system, is an order of succession, therefore from his perspective mathematical series also represent a developmental regularity analogous to the order of time, that is analogous to the successions of perceptual states of thinking beings. Distinct and heightened perceptions are analogous to the noticed members of a mathematical series (e.g. 1,4,9,16, 25) and the hidden infinitesimal regularity governing the sequence is analogous to the most determined path from one perceptual state to the other representing the order of time of the best possible world.[25] Or in Leibniz's terms: "I had said that temporal events follow from particular things [so that] if I am not mistaken, an

---

[22] Leibniz, "Letter to Varignon," 1702, *G.W. Leibniz: Philosophical Papers and Letters*, p. 544.

[23] Leibniz, "Letter to Varignon," 1702, *G.W. Leibniz: Philosophical Papers and Letters*, p. 544. My emphasis.

[24] Leibniz, "The Principles of Nature and of Grace, Based on Reason," & 13, 1714, *Die Philosophischen Schriften von G.W. Leibniz*, vol. 4, pp. 598-606, translated in: *G.W. Leibniz: Philosophical Papers and Letters*, p. 640. My emphasis.

[25] This issue will be further developed below.

essential order of particulars corresponds to the definite parts of time and space."[26]

To further substantiate my argument that in Leibniz's system operation of differentiation functions as a rational technique assisting human beings to discover the hidden law of the generation of curves, I would like to introduce a mystical text from his Paris Notes (1676). Most of the notes Leibniz wrote in these creative moments ought to be read cautiously since he set them down in a brainstorming disposition and later modified many of his ideas. Nevertheless, I believe that the following description of his spiritual experience of the "mind's self-reflection" is a key to understanding what lies behind the operations of the calculus, discovered during the same years. He writes thus:

> God is the perfect mind, and that mind is the cause of its own perceptions; which is not the case with any other mind.... The following operation of the [human] mind seems to me to be most wonderful: namely that when I think that I am thinking, and in the middle of my thinking I note the I am thinking, and a little later I wonder at this tripling of reflection. Next I also notice that I am wondering and in some way I wonder at this wonder, and fixed in one contemplation I return more and more into myself, alternately as it were, and often *elevate my mind through my thoughts*."[27]

Let me interrupt the text here for a moment in order to draw the reader's attention to the analogy between Leibniz's description of self-reflexivity and the operation of differentiation. The first reflection of the mind exposes Leibniz to a finite sequence of his train of thoughts that seems to have no noticed regularity, the second reflection upon the reflection is analogous to the operation of the first order of differentiation, the next reflection is already a second-order differentiation, and so on. Also as the result of self-reflexivity the mind becomes elevated and the thoughts are refined,[28] as the third-order of differences of the finite square sequence (1, 4, 9, 16, 25) becomes nullified. Furthermore, during the refining spiritual process the elevated mind is able to grasp its hidden lawfulness in a way similar to the way mathematicians uncover the hidden algorithm of the curve through the process of differentiation. Let us return now to the text:

---

[26] Leibniz, "Correspondence with De Volder," June 30, 1704, *Die Philosophischen Schriften von G.W. Leibniz*, vol. 2, pp. 270-1, translated in: *G.W. Leibniz: Philosophical Papers and Letters*, p. 538.

[27] Leibniz, "On Reminiscence and on the Mind's Self-Reflection," 1676, Garber and Sleigh (eds.), *De Summa Rerum: Metaphysical Papers*, pp. 71-3, my emphasis.

[28] I believe this is what he means when he says later in the text that "we are brought back within ourselves and suppress what is external." "On Reminiscence and on the Mind's Self-Reflection," p. 73.

> For this [the elevation of thought] to come about it is necessary that the power and vivacity of other extraneous perceptions shall be weakened and broken by some effort of the nerves.

Once again, the analogy that comes to mind is that differentiation focuses the attention on the difference between the terms of the sequence so that the result is similar to the one mentioned above - the finite sequence fades from our attention while we begin to grasp its hidden regularity. After a short detour into the "difficulty of sleeping," Leibniz continues:

> I have noticed, that this perception of perceptions also occurs without symbols.... I have not yet explained satisfactorily how there comes about these different beats of the mind, with that constantly reciprocated reflection, and, as it were, the intervals of these beats.[29]

He continues:

> They [the beats] seem to occur by the distinguishing awareness of the corporeal intention; but, if you observe carefully, the beat only brings it about that you remember that you had this – namely the reflection of a reflection- in the mind a little before, and you, as it were, observe this, and designate it by a distinct image which accompanies it. Therefore it already existed before, and *so the perception of a perception to infinity is perpetually in the mind, and in that consists its existence per se, and the necessity of its continuation.*[30]

This wonderful text focuses our attention on the trivial truth of self-reflexivity, yet includes remarkable insights into the nature of souls, about which Leibniz's mature *Monadology* has much to say. But what is most remarkable, for our discussion, is the insight this text gives on the possible role of higher differentials in unraveling the order of time that is rolled up and concealed in the confused perceptions of monads. The issue of the order of time and the deeper implications of Leibniz's calculus to the *Monadology* will be elaborated below. For the time being it is enough to accept that Leibniz maintained that temporal successions are also a series. He says this explicitly to Burcher De Volder (1643-1709):

---

[29] To my understanding, Leibniz is reflecting here upon a similar phenomenon that is described in the calculus, which he called the indeterminacy of differentials.

[30] Leibniz, "On Reminiscence and on the Mind's Self-Reflection," pp. 73-5. My emphasis.

> You say that in a series, such as one of numbers, nothing is thought of as a successive. What of it? I do not say that every series is a temporal succession but only that a temporal succession is a series, which has in common with other series the property that the law of the series shows where it must arrive in continuing its progress, or in other words, the order in which its terms will proceed when its beginning and the law of its progression are given, whether the order is a priority of essences only or also one of time.[31]

The contrast between Newton's and Leibniz's mathematical insight is now apparent. Newton studied the fluxion of quantities in relation to the equable flow of absolute time. For him, quantities flow within time while generating a continuum of values. A curve indicates that a rate of change has caused a deviation in the otherwise continuous equable flowing quantity; therefore the Newtonian mathematician will study the deviation of the curve from an external preferred equably flowing quantity. In contrast, a Leibnizian mathematician will see the curve as a finite limited expression of a hidden divine infinitesimal developmental regularity. The curve represents a mathematical series and not a continuous fluxion. Differentials are variables ranging over number sequences and geometrical curves. Finite sequences and curves are investigated from an internal perspective (see e.g. the characteristic differential triangle), which considers them to be a partial expression of a more fundamental hidden differential regularity. Thus Leibniz's operation of differentiation generates an additional set of variables which range over another infinite sequence of infinitely near values and are not, as in Newton's fluxions, part of an external frame of reference associated with a uniformly and continuous flowing quantity. Put differently, Leibniz's differentials are internally contained in the sequence and when operated they generate another sequence which is a more distinct infinitesimal image of the original finite sequence, whereas Newton's chosen uniformly flowing quantities do not belong to the changing fluxion analyzed and are only an external device for calculating the rates of change these deviated fluxions undergo in time.

Further, in contrast to Leibniz's definition of the tangent as a special kind of differential, Newton interpreted the tangent as a most informative enduring frame of reference. For Newton, in each curvature trajectory of a physical body, tangents express the inertial flow of the body if nothing has disrupted its *vis insita* at the investigated moment. Thus, tangents are chosen fluxions flowing along the equable flow of God's endurance, on whose basis one ought to investigate the deviated trajectory caused by central forces.

---

[31] Leibniz, "Correspondence with De Volder," January 21, 1704, *Die Philosophischen Schriften von G.W. Leibniz*, vol. 2, pp. 264-5, translated in: *G.W. Leibniz: Philosophical Papers and Letters*, p. 534.

Tangents are informative devices since they recover the equable flow of God's absolute time. Going back to Leibniz's interpretation of the tangent, we have a totally different account. The Leibnizian differential, though also a device, does not represent an equable flow of time. On the contrary, the differentials representing the tangents of the original curve (the finite sequence) unfold the dynamic progression of the consecutive terms of the sequence (e.g. in the above example: 0, 1, 4, 9, 25).

So far I have discussed the difference in the capacities and operations of the two calculi. To consider the structural resemblance of Newton's method of fluxions to his physical, theological and historical work we shall look at his understanding of God's infinite perspective, then compare it to Leibniz's insights, and return to the method of fluxions and the calculus in order to further define their role in the respective wider systems. To delineate the role of the two mathematical methods in achieving a God-like perspective I shall first analyze Newton's and Leibniz's opposing notions of space and time, which in both systems are very closely connected to God's perspective.

CHAPTER V

NEWTON'S AND LEIBNIZ'S NOTIONS OF SPACE AND TIME

> "Absolute, true, and mathematical time, of itself, and from its own nature flows equably without regard to anything external, and by another name is called duration, [whereas,].... absolute space, without regard to anything external, remains similar and immovable." (Newton, Principia, p. 13).

> "Space, even as time, is nothing other than an order of possible existences, simultaneously in the case of space, or successively in the case of time, and in themselves have no reality." (Leibniz, *Leibniz: Mathematische Schriften*, vol. 7, p. 242).

I. NEWTON

In the *Principia*, Newton defines absolute time as prior to absolute space and assigns it more structure. The first scholium of the *Principia* reads:

> Absolute, true, and mathematical time, of itself, and from its own nature flows equably without regard to anything external, and by another name is called duration, [whereas,].... absolute space, without regard to anything external, remains similar and immovable.[1]

These definitions and the text that follows them make clear that absolute space and absolute time also condition the physical qualities of material bodies, especially their quantitative character.[2] Without the existence of these absolutes the quantifying project of the *Principia* (of physical notions such as mass, velocity, *vis insita*, gravitational force, etc.) would not have been possible. Space and time function as the objective measuring rods for all physical magnitudes.

Arthur points out that Newton insisted that space is prior to material

[1] *Principia*, p.13.

[2] See for example all of the first eight definitions of the *Principia*, pp. 9-13, and Cohen's clear exposition on these concepts in: Cohen-Whitman, *Isaac Newton: The Principia*, pp. 106-108.

bodies.[3] The best explicit support for this thesis is given in the earlier *De Gravitatione*. After presenting space and time, Newton adds a long hypothetical argument, saying: let's assume that "God has happened to endow various delimited regions of empty space with the following three conditions: (i) mobility, (ii) impenetrability (so that on collision they reflect according to the laws of motion), and (iii) the ability to excite the various perceptions of the senses and the faculty of fancy in created minds." He concludes that under these three conditions specified spaces would be indistinguishable from material bodies.[4]

More important, without the absolute external frames of reference the innate passive force, *vis insita*, of atoms would also be meaningless because it would not be clear in respect to what bodies remain in an inertial flow. The equable flow of absolute time thus conditions the innate endeavor of all material atoms to maintain their physical state in correlation to time's inertial flow. The *vis insita* of bodies preserves an inertial flow, though the atom itself is constantly disrupted by external interactions which keep changing its rate of flow.

Furthermore, the structure of time also very subtly conditions the individuation of matter into separate and non-destructible atoms. The reasoning is given implicitly in the *Principia*:

> The duration or perseverance of the existence of things remains the same, whether the motions are swift or slow, or none at all.[5]

The structure of absolute time does not permit existential changes in the perseverance of things. Thus atoms are non-destructible and non-penetrable entities, and the *vis insita* of each atom constantly perseveres its inertial flow (constantly maintaining it even when disrupted). More precisely, Newton believes that most observed phenomenal changes are due to external disruptions of the equable rate of flow of non-destructible entities, and that *vis insita* is constantly at work to repair all disruptive changes (such as accelerations). Nonetheless, disruptions constantly occur in the physical state

---

[3] Richard Arthur, "Space and Relativity in Newton and Leibniz," *The British Journal of Philosophy of Science* vol. 45 (1994), pp. 219-40, p. 229.

[4] Newton, *The Correspondence of Isaac Newton*, A.R. Hall and M.B. Hall (eds.), (Cambridge, 1962) pp. 106-9, 140-2; Richard T. W. Arthur, "Newton's Fluxions and Equably Flowing Time," p. 229. For a fuller description on the definitions of space and time in *De Gravitatione*, see: McGuire, "Existence, Actuality and Necessity: Newton on Space and Time," *Annals of Science*, vol. 35 (1978), pp. 463-508; McGuire, "Space, Infinity, and Indivisibilty: Newton on the Creation of Matter," Zev Bechler (ed.), *Contemporary Newtonian Research* (Dordrecht, 1982), pp. 152-3; McGuire, "The Fate of the Date: The Theology of Newton's Principia Revisited," Margaret J. Osler (ed.), *Rethinking the Scientific Revolution*, (Cambridge, 2000), pp. 279-283.

[5] *Principia*, p.15.

of atoms. However, even here the structure of the absolutes conditions the nature of the action as well as the reaction of each minute atom to any interference with its equably flowing state.[6] Newtonian time and space function as the external limiting cases conditioning the third law of the motion of atoms, assuring that "to every action there is always opposed an equal reaction."[7] Newton's third law of motion is already contained in *vis insita*'s dual reaction (of resistance and impulse) to any interference with its inertial state:

> But a body exerts [*vis insita*] only, when another force, impressed upon it, endeavors to change its condition; the exercise of this force may be considered both as resistance and impulse; it is resistance, in so far as the body, for maintaining its present state, withstands the force impressed; it is impulse, in so far as the body, by not easily giving way to the impressed force of another, endeavors to change the state of that other.[8]

All atoms residing in the absolute containers are constantly at work to cancel any disruption of the equable flow of the whole material system.[9] For this reason, "whatever draws or presses another is as much drawn or pressed by that other."[10] Thus the containers themselves, especially time's equable flow, become the limiting case conditioning the laws of physical interactions. Further, the similarity of the structure of equable flowing absolute time and the *vis insita* of atoms, absolves bodies of the need for an internal organization or memory since once a certain state has been acquired it will be preserved. To reiterate, nothing happens in the equable flowing duration, so that no unfolding process needs to be memorized or conserved in order to preserve that preferred state. Physical interactions and phenomenal changes are treated as external (non-essential) disruptions, against which the unchanging atom reacts. Thus Newton's atoms do not need an internal organization, which unfolds as they develop in time; the atom *per se* undergoes no changes, it is non-destructible. All it needs is a protective passive *vis insita*, which constantly restores the lost equable flow caused by external interactions. Any kind of order or organization found in nature or in bodies cannot be due to the internal passive force of matter but has to come from an external source, such as gravity or even from

---

[6] On this issue see also: Funkenstein, *Theology and the Scientific Imagination*, p. 94, where he says that "Absolute space and God are both preconditions for actions."

[7] *Principia*, p. 19.

[8] *Principia*, p. 10, cited also previously in chapter III.

[9] Newton: "... the common centre of gravity of all bodies acting upon each other (excluding outward actions and impediments) is either at rest, or moves uniformly in a right line." *Principia*, Corollary IV, p. 23.

[10] *Principia*, p. 19.

God himself. Put differently, in Newton's system active forces that shape and organize passive matter into noticed structures are never material or internal to matter, but have a different active source. Matter is always passive. All it is programmed to do is to protect its momentary inertial state according to the three laws of motions and in this respect it has a built in remembrance of the equable flow of God's absolute time within its massive fabrics.

II. LEIBNIZ

Going on to Leibniz's notions of space and time, we encounter a totally different theory. His fundamental entities are not passive material atoms residing within absolute containers, but active spiritual entities endowed with a perceptual internal developmental regularity. Space and time do not exist *per se* as containers; instead they are the orders of existence of things. Space is the order of coexisting things and time is the order of succession of things.[11] The order of succession is a developmental order of states, it is the regularity according to which a specific state passes into the succeeding one and the unfolding of the whole ensuing series of states. Or in Leibniz's terms:

> But space and time taken together constitute the order of possibilities of the one entire universe, so that these orders – space and time, that is – relate not only to what actually is but also to anything that could be put in its place, just as numbers are indifferent to the things which can be enumerated.[12]

In Leibniz's system we do not have enduring states like Newton's equable flow, since time is not a measure of duration but the order of possible transitions from one state to the one following it. "For *space*," writes Leibniz in a letter to De Volder, "is nothing but the order of existence of things possible at the same time, while *time* is the order of existence of things possible successively."[13] Thus space and time are not God's absolutes (as in Newton),

---

[11] Leibniz, "The Metaphysical Foundations of Mathematics," written after 1714, *G.W. Leibniz: Philosophical Papers and Letters*, p. 666.

[12] Leibniz, "Reply to the Thoughts on the System of Pre-established Harmony Contained in the Second Edition of Mr. Bayle's Critical Dictionary, Article Rorarius," 1702, *Die Philosophischen Schriften von G.W. Leibniz*, vol. 4, pp. 554-71, translated in: *G.W. Leibniz: Philosophical Papers and Letters*, p. 583.

[13] Leibniz, "Letter to De Volder," June 1704, *Die Philosophischen Schriften von G.W. Leibniz*, vol. 2, pp. 268-71, translated in: *G.W. Leibniz: Philosophical Papers and Letters*, p. 536. See also "Time is the order of existence of those things which are not simultaneous. Thus time is the universal order of changes... Space is the order of coexisting things." Leibniz, "The Metaphysical Foundations of

but possible orders of existence of possible worlds:

> Space, even as time, is nothing other than an order of possible existences, simultaneously in the case of space, or successively in the case of time, and in themselves have no reality.[14]

Leibniz's God perceives infinitely many possible worlds differing in their orders of succession and coexistence. He analyzes all these possible orders of coexistence and succession and chooses to create only the world that has the best order of existence.[15] Leibnizian time is not a Newtonian equable flowing state, but an ongoing transition of states. An equable flow is an expression of a body that remains the same through all its transitional states; such a possible chain of events is not possible in Leibniz's world, since everything there develops continuously according to a most determined regularity. Further, Leibniz's space, as an order of coexistence of simultaneous states, is never similar to itself since its differing successive states keep changing, nor is it immovable, because it does not exist *per-se* as Newton's absolutes. On the contrary, the order of coexistence of any possible world contains in a potential form the preliminary and ensuing developmental successive regularity of all coexisting entities. As Arthur says, "a monad is a *fibre bundle*: in addition to the one-dimensional manifold of states, we have for every state something analogous to the derivative, a monadic state of change, which is, formally, a tangent space so that, mathematically speaking, the monad is a *tangent bundle*."[16]

Leibniz distinguishes between two differing phases of time and space as orders of existence. In the first, space and time are ideal orders in the abstract prior to any actualization of things; in the latter, space and time are the orders of existence of possible worlds. The first, more abstract, space and time are only a capacity for ordering prior to any realization of specific things and does not concern us here, whereas the latter concerns concrete orders of existence of possible worlds.[17] Space and time as concrete orders of existence have many possible different realizations. Nonetheless, "out of the infinite combinations and series of possible things, one exists through which the

---

Mathematics," *G.W. Leibniz: Philosophical Papers and Letters*, p. 666, see also: pp. 531, 656.

[14] Leibniz, *G.W Leibniz: Mathematische Schriften*, vol. 7, p. 242 translated by Glenn A. Hartz and J.A. Cover, "Space and Time in the Leibnizian Metaphysic," *Nous* vol. 22 (1988), pp. 493-519.

[15] *G.W. Leibniz: Philosophical Papers and Letters*, p. 583.

[16] Richard Arthur, "Space and Relativity in Newton and Leibniz," p. 229, his emphasis.

[17] On this issue see F.S.C. Northrop, "Leibniz's Theory of Space," *Journal of the History of Ideas* vol. 8 (1946), pp. 422-446; A.T. Winterbourne, "On the Metaphysics of Leibnizian Space and Time," *Studies in the History and Philosophy of Science*, vol. 13 (1982), pp. 201-14; Richard T.W. Arthur, "Leibniz's Theory of Time," K. Okruhlik and J.R. Brown (eds.), *The Natural Philosophy of Leibniz* (Reidel, 1985), pp. 263-313.

greatest amount of essence or possibility is brought into existence,"[18] and the best order of existence among possible worlds is our created world. But what does it actually mean that a world possesses the best order of existence? Leibnizian scholars are not in agreement on this issue. For my purposes, I will adopt an interpretation,[19] which argues that the best choice is the one which prefers the maximum effect at the minimum cost. When dealing with orders of existence this means achieving maximum coexistence among diverse entities with the investment of a minimal internal successive regularity governing these entities. Thus the chosen spatial and temporal orders of our world condition a world populated with a maximal number of entities, each unfolding according to a minimal and "most determined" successive regularity (lawfulness), which guarantees the harmonious coexistence of all ensuing states of development.[20] To illustrate the meaning of the best choice in terms of order of existence Leibniz gives us the following example:

> Hence it is clearly understood that out of the infinite combinations and series of possible things, one exists through which the greatest amount of essence or possibility is brought into existence. There is always a principle of determination in nature which must be sought in by maxima and minima; namely that a maximum effect should be achieved with a minimum outlay, so to speak. And at this point time and place or in a word, the receptivity or capacity to the world, can be taken for the outlay, or the terrain on which a building is to be erected as commodiously as possible, the variety of forms corresponding to the spaciousness of the building and the number and elegance of its chambers.[21]

In other words, "there is a definite rule by which a maximum number of spaces can be filled in the easiest way." The following is an example from the calculus for achieving a maximum outcome with minimum expenditure: "assuming that there is to be a triangle with no further determining principle, the result is that an equilateral triangle is produced. And assuming that there is to be motion from one point to another without anything more determining the

---

[18] Leibniz, "On the Radical Origination of Things," *G.W. Leibniz: Philosophical Papers and Letters*, p. 487.

[19] For a similar attitude see: David Blumenfeld, "Leibniz's Theory of the Striving Possibilities," *Studia Leibnitiana* vol. 5 (1973), pp. 163-77; Donald Rutherford, *Leibniz and the Rational Order* (Cambridge, 1995).

[20] Leibniz, "On the Radical Origination of Things," *G.W. Leibniz: Philosophical Papers and Letters*, p. 487.

[21] Leibniz, "On the Radical Origination of Things," 1697, *G.W. Leibniz: Philosophical Papers and Letters*, p. 487.

route, that path will be chosen which is easiest or shortest."[22] The triangle exemplifies the choice of the best order of coexistence (space), since the equilateral triangle has the greatest area in comparison to all other triangles having the same perimeter (and as such it expresses the greatest possible order of coexistence since it can be populated by the greatest amount of entities in a given space); whereas the path from one point to the other one exemplifies the best order of succession (time) since the easiest and shortest path is the simplest regularity that can govern any given series of states. Therefore, the best orders of existence of a possible world will be those governed according to the above principle of determination.

This principle brings us back to the major difference between Newton and Leibniz as already introduced in chapter two regarding the design of the world. According to Newton, the differential equations have nothing to say regarding the initial conditions of the design God chose to create, whereas Leibniz says explicitly in this text and others that even the initial conditions of the best possible world are determined according to the principle of determination, or to put it a bit differently:

> This principle of nature, that it acts in the most determined ways in which we may use, is purely architectonic in fact, yet it never fails to be observed. [He gives the same example of an equilateral triangle as given above, and concludes:] This example shows the difference between architectonic and geometric determinations. Geometric determinations introduce an absolute necessity, the contrary of which implies a contradiction, but architectonic determinations introduce only a necessity of choice whose contrary means imperfection – a little like the saying in jurisprudence: "things which are contrary to moral principles, we ought also to believe we are unable to do." So there is even in the algebraic calculus what I call the law of justice, which greatly aids in finding good solutions.[23]

Thus we see that Leibniz gave his calculus an architectonic meaning and role in the affairs of the world, which he associated with the law of justice. This means that he associated rationality and the calculus with moral laws. The best choice is always the most rational one and can be calculated with the aid of the calculus. The calculus enables men to grasp God's choice of creating the best possible orders of existence (space and time) though no man can know the infinite as distinctly as God does. The calculus and the architectonic laws it

---

[22] Leibniz, "On the Radical Origination of Things," 1697, *G.W. Leibniz: Philosophical Papers and Letters*, p. 487.

[23] Leibniz, "Tentamen Anagogicum: An Anagogical Essay in the Investigation of Causes," 1696, *G.W. Leibniz: Philosophical Papers and Letters*, p. 484.

exposes assure men that truth, beauty, and the good are all connected in God's rational mind. In addition, nature (the creation) is also "governed architectonically" and the "half-determinations of geometry are sufficient for it to achieve its work."[24] The general law of justice assures men that God chose the best and most rational initial conditions (as in the case of the equilateral triangle) for our world.

Leibniz's spatial and temporal orders are mutually dependent upon one another and both are governed by the best choice. The spatial order potentially contains the temporal order, and vice versa; the successive order unfolds the order of coexistence. Moreover, Leibnizian space and time are orders of things (in contrast to the Newtonian absolutes) conditioned by a choice. They govern the individuation of entities only indirectly via the preference for the best order. Let us go further into the meaning of the notion that the spatial and temporal orders of our world are the best orders of existence. The choice of these orders has an immediate consequence upon the individuation of entities and the spectrum of possible operations of things in the created world. The orders of existence of a world condition also the laws of nature that will govern material bodies. Laws of nature limit the possible degrees of freedom of a physical system since they restrict the possible outcomes of interactions. In Leibniz's system they also condition physical forces to have a metaphysical inner source.

The reason for this is that the best order of existence conditions all coexisting entities to have internal active forces, which regulate their infinitesimal and finite development according to the chosen determined pre-established harmony of coexistence and succession. In his mature philosophy, Leibniz calls the ultimate units of reality monads.[25] Monads are not material atoms but spiritual beings possessing perceptions and appetitions, each expressing the world from a particular point of view. The best order of existence conditions the makeup of monads. Monads express intrinsically the world they inhabit since the content of each transient state of a monad necessarily includes and is regulated by the best choice. Thus each perceptual state of a monad (including many unnoticed perceptions) expresses (with differing degrees of clarity) maximal coexistence (the spatial order) developing according to a minimal successive regularity (the temporal order).[26] In contrast to Newton's atoms, which have no internal memory of the world they live in, Leibniz's monads express their world according to an internal developmental organization. Whereas Newton's absolutes function as external limiting cases conditioning the material world, Leibniz's temporal and spatial orders are programmed internally into the fabrics of each individual monad, as suggested in the text below:

---

[24] *G.W. Leibniz: Philosophical Papers and Letters*, p. 484.

[25] The most explicit text on monads is: "The Monadology," (1714), *G.W. Leibniz: Philosophical Papers and Letters*, pp. 643-654.

[26] Leibniz, "The Monadology," and "The Metaphysical Foundations of Mathematics," *G.W. Leibniz: Philosophical Papers and Letters*, p. 666.

> The essential ordering of individuals, that is, their relation to time and place, must be understood from the relation they bear to those things contained in time and place, both nearby and far, a relation which must necessarily be expressed by every individual, so that a reader could read the universe in it if he were infinitely sharp-sighted.[27]

According to Newton, the general laws of motion governing bodies only determine that atoms react out of a persevering passive force if they are disrupted according to the three laws of motion. This means that once an interaction has disrupted the inertial state of an ultimate constituent of reality, its *vis insita* is predetermined to react in order to establish another equably flowing state. The essence of Newton's atoms does not include the environment, but only a reminiscence of time's endurance. The environment is external to atoms and bodies and influences the body only when an interaction occurs; it is then considered an obstacle and distraction to the atoms' enduring existence. Thus in a metaphorical sense, when entering into a public interaction, atoms care only for their self-preservation, ignoring the consequences that the interaction has upon the whole system. The laws of motion restrict the degrees of freedom of bodies only to a certain degree. Much freedom still remains for things to go wrong in regard to the original design of creation since there are no laws of conservation, and motion is lost in most interactions. For this reason Newton's God, as the governor, needs to amend the system from time to time. Without God's constant sustenance and periodic amendment matter alone will never form organized structures.

In contrast, Leibniz's God and his creatures are constantly restricted to make the best moral choice and to maximize existence at each and every step they take. Once God had chosen the most determined and minimal order of succession (time), the progression of all the states of the created world, including all individuals' perceptual states, necessarily develop according to that rule.[28] Whereas Newton's laws of nature left room for providence to intervene when too great a disorder entered the system, Leibniz's "half-determined architectonic" laws limited and restricted all future events to such a degree that nothing happened without a reason governed according to the best choice. In the Leibnizian world everything is related to everything else, there are no impenetrable Newtonian atoms separated from one another. The environment and all its intricate inter relations is the internal perceptual reality

---

[27] "Leibniz to De Volder, (1704/1705)," *G.W. Leibniz: Philosophical Essays*, p. 183.

[28] "There is, moreover, a definite order in the transition of our perceptions when we pass from one to the other through intervening ones. This order, too, we call a *path*. But since it can vary in infinite ways, we must necessarily conceive of one that is most simple, in which the order of proceeding through determinate intermediate states follows from the nature of the thing itself, that is the intermediate stages are related in the simplest to both extremes." Leibniz, "The Metaphysical Foundation of Mathematics," *G.W. Leibniz: Philosophical Papers and Letters*, p. 671.

of individual monads. This demand for harmonious coexistence is what restricts the freedom of motions in the Leibnizian world. Whereas Newton insisted on the self-perseverance of atoms, thus giving autonomy to the parts of the system to seek their own endurance, so to speak, Leibniz's best order of succession controls the whole as well as the development of all the infinitesimal parts. In such a holistic order of successions (time) there is no place for an egoistic/free choice that does not follow the highest and most rational and moral code, since the order of succession of the entire world is determined to maximize harmonious coexistence.

To sum up: Newton's absolute space and time, by existing *per se* and flowing equably, condition the three laws of motion of the material reality and the composition of non-destructible atoms. The absolutes function as external limiting cases of physical reality. The laws of motion have nothing to say regarding the original design God imprinted upon matter. All they do is assure us that there is no action of bodies in the world that does not have a reciprocal reaction. Leibniz's spatial and temporal orders are orders of existence of metaphysical entities. They do not exist independently of things. They are possible orders of existence of things. Thus once the best order of existence is chosen the spatial and temporal orders are programmed internally into the fabrics of monads. Leibniz's space and time do not function as external limiting cases of phenomenal reality but as the hidden minimal internal developmental regularity, which enables a maximal number of monadic entities to coexist harmoniously. Leibniz's architectonic laws of nature leave no room for free will in the Newtonian sense. They limit and restrict the degree of freedom of physical systems to such an extent that nothing happens without a reason.

CHAPTER VI

GOD'S ABSOLUTE PERSPECTIVE ACCORDING TO NEWTON

"Accordingly those who interpret [space and time] as referring to the quantities being measured do violence to the Scriptures. And they no less corrupt mathematics and philosophy who confuse true quantities with their relations and common measures." (Newton, *Isaac Newton: The Principia*, p. 414).

I. THE PRISTINE ASPECTS OF THE ABSOLUTES

It is clear from Newton's writings that he maintained humans could never know which uniformly flowing quantities really correlate with absolute time's equable flow.[1] However, he conceded that all uniformly flowing quantities belong to a preferable group, which in a finite way exposes the uniform flow of divine mathematical time. In what follows I will argue that we have good reason to assume that Newton's association of fluxions with God's absolutes, together with the project of the *Principia* is a modern form of the ancient religious worship of Noah and his sons around vestal fires, incorporating the mathematical concepts inherent in the ceremony reflecting the motion of the planets around the sun.[2]

Newton maintained that human beings' bodies are part of the physical world. This means that the third law of motion of the reciprocity of action and reaction also governs people's bodily perceptions once they engage in material interactions. In most cases, mortals cannot achieve a truly divine absolute perspective, since they are always involved with the interactions they try to understand. All the impressions that reach their sensorium are mediated and

---

[1] *Principia*, Book I, Scholium I, especially pp. 15-16. See also my paper: Ayval Ramati, "The Hidden Truth of Creation: Newton's Method of Fluxions," *The British Journal for History of Science*, vol. 34 (2001), pp. 417-438, pp. 429-431.

[2] Dobbs, Snobelen, Rudolf De Smet and Karin Verelst suggest a similar idea, as will be discussed below. Dobbs suggests that Newton "saw his achievement not as something new under the sun but rather as a step toward the restoration of the true natural philosophy of the ancients." *Janus Faces,* p. 210. Snobelen points out that Newton's theological ideas appear in an esoteric form in the "General Scholium," see his: "God of gods," pp. 202-208. Rudolf De Smet and Karin Verelst point out that the theological portion of the General Scholium reveals a classical strata. See: Rudolf De Smet and Karin Verelst, "Newton's Scholium Generale: the Platonic and Stoic legacy — Philo, Justus Lipsius and the Cambridge Platonists," *History of Science* vol. 39 (2001), pp. 1-30.

interfered with since they are part of a segment of a chain of actions and reactions of the physical world by which they themselves are influenced without even being aware of it. Whatever they choose to do they will always be acted upon in return. Consequently, mortals' perceptions will most likely deviate from God's absolute equably flowing perspective. It is a human predicament to perceive interactions in the world in a distorted way through an action-reaction mechanism, unless they understand and accord with God's physical and moral laws.[3] According to the *Principia*, "the vulgar," who have not understood the general laws of motion will "conceive those quantities [time, space, place, and motion] under no other notions but from the relation they bear to sensible objects. And thence arise certain prejudices, for the removing of which, it will be convenient to distinguish them into absolute and relative, true and apparent, mathematical and common."[4] The *Principia*, on the other hand, offers epistemological methods to assist humans in adjusting and amending their distorted sensible perception of reality with the aid of mathematics and an understanding of the nature of the true absolutes (space and time). Newton compares the confusion of the vulgar with the erroneous interpretation of the sacred writings. At the end of the scholium on the absolutes, he writes:

> Wherefore relative quantities are not the quantities themselves, whose name they bear, but those sensible measures of them (either accurate or inaccurate), which are commonly used instead of the measured quantities themselves. And if the meaning of words is to be determined by their use, then by the names time, space, place, motion, there measures are properly to be understood; and the expression will be unusual, and purely mathematical, if the measured quantities themselves are meant. Upon which account, they do strain the sacred writings, [those] who interpret those words for the measured quantities. Nor do they less defile the purity of mathematical and philosophical truths, who confound real quantities themselves with their relations and vulgar measures.[5]

---

[3] Snobelen correctly suggests that Newton's powerful Monarchian view of God did not allow neither evil spirits nor Satan himself to deceive the senses. Therefore "Newton was not faced with Descartes' demon, who could distort our perception of reality and thus call into question the results of experiments in fields such as optics. God's universal and unchallenged dominion made such malevolent deception of the senses impossible." Snobelen, "To Discourse of God: Isaac Newton's Heterodox Theology and his Natural Philosophy," forthcoming. In the discussion that follows, I add to this insightful observation a correction and an explanation of a divine mechanism of sensation (the sensorium) may become corrupted through a vulgar misuse.

[4] Newton, *Principia*, first scholium, p. 13.

[5] Newton, *Principia*, p. 17-8. Cohen and Whitman translate this text as follows: "Relative quantities, therefore, are not actual quantities whose names they bear but are those sensible measures of

The vulgar misconception of physical reality is damaging. It is a distorted perception arising "from the relation [human beings] bear to sensible objects."[6] Instead, absolute notions arise from the pure mathematical relation humans have with God's absolute duration. Thus "absolute time, in astronomy, is distinguished from relative, by the equation or correction of the vulgar time."[7] The analogy between vulgar and relative notions and the corruption of the original religion through idolatry may be introduced at this early point. In both idolatry and vulgar misconceptions of physical quantities the immediate objects which human beings relate to are sensible objects. The consequence of not connecting with God alone or with his absolutes is distortion and corruption of perceptual reality, of morality, and of the understanding of human beings.[8] Newton's use of such strong words such as strain and defile, in the above text, to describe the vulgar notions of space and time in contrast to the use of the notion of purity to describe mathematics dealing with the absolutes is telling. The following text takes this idea further:

> It has been necessary to distinguish absolute and relative quantities carefully from each other because all phenomena may depend on absolute quantities, but ordinary people who do not know how to abstract their thoughts from the senses always speak of relative quantities, to such an extent that it would be absurd for either scholars or even Prophets to speak otherwise in relation to them. Thus both the Sacred Scripture and the writings of Theologians must always be understood as referring to relative quantities, and a person would be labouring under a crass prejudice if on this basis he stirred up arguments about absolute [*changed to* philosophical] notions of natural things.[9]

---

them (whether true or erroneous) they are commonly used instead of the quantities being measured. But if the meanings of words are to be defined by usage, then it is these sensible measures which should properly be understood by the terms "time," "space," "place," and "motion," and the manner of expression will be out of ordinary and purely mathematical if the quantities being measured are understood here. Accordingly those who interpret these words as referring to the quantities being measured do violence to the Scriptures. And they no less corrupt mathematics and philosophy who confuse true quantities with their relations and common measures." Cohen/Whitman, *Isaac Newton: The Principia*, scholium on space and time, pp. 413-4.

[6] Newton, *Principia*, p. 13.

[7] Newton, *Principia*, p. 14.

[8] Indeed, Manuel has observed that Newton found a strong link between the growth of idolatrous polytheism (including the Trinity) and the corruption of natural philosophy, see: Manuel, *The Religion of Newton*, p. 42. Dobbs also demonstrated the ways in which Newton's antitrinitarian theology is related to his philosophy of nature in: Dobbs, *Janus Faces*, pp. 213-49. See also: Snobelen, "To Discourse of God."

[9] Newton, *Isaac Newton, The Principia*, p. 36. See also: Snobelen, "God of gods," pp. 204-208.

The reason behind this imaging of relative and vulgar concepts versus the pure mathematical relation with God is further explained in one of the queries of the *Opticks* where Newton introduces God's sensorium as synonymous to absolute space. God "being in all Places, is more able by his Will to move the Bodies within his boundless uniform Sensorium." More remarkable is the fact that God has endowed human beings with a soul similar to his infinite, eternal sensorium:

> [God] is a uniform Being, void of Organs, Members or Parts,... and he is no more the Soul of [his creatures], than the Soul of Man is the Soul of the Species of Things carried through the Organs of Sense into the place of its Sensation, where it perceives them by means of its immediate Presence, without the Intervention of any third thing. The Organs of Sense are not for enabling the Soul to perceive the Species of things in its Sensorium, but only for conveying them thither; and God has no need of such Organs, he being every where present to the Things themselves.[10]

For Newton, writes Dobbs, "as we sense objects indirectly when their images are brought home to our little sensories, the centers of perception in our brains, so God perceives objects directly."[11] Newton states this clearly when he says that God "has no need for such Organs" since God perceives everything directly (with no mediation) through his infinite sensorium. "He endures forever, and is everywhere present; and by existing always and every where, he constitutes space and time."[12] Yet we mortals need "organs of sense" to mediate between the objects outside our body and our sensorium. It is here that confusion, error, and corruption may enter. More to the point, as long as human beings are not aware of the intricate mechanism of their sensorium and their sense organs and do not understand that according to the three laws of motion the images brought into their sensorium may have been distorted due to material interactions they will mistakenly construct relative and vulgar notions of space and time.

Yet the *Principia* offers humanity a way out of this vulgar fallacy once the mathematical principles of reality are fully understood. Newton thought he had not invented anything new in his mathematical analysis, since this truth was already known to the ancients:[13]

---

[10] *Opticks*, query 31, p. 403.

[11] Dobbs, "Newton's Alchemy and his 'Active Principle' of Gravitation," P.B. Scheurer and G.Debrock (eds.), *Newton's Scientific and Philosophical Legacy* (Dordrecht, 1988), p. 73.

[12] *Principia*, General Scholium, p. 441.

[13] See also: P. M. Rattansi, "Newton and the wisdom of the ancients," J.Fauvel, Raymond Flood, Michael Shortland and Robert Wilson (eds.), *Let Newton be! A New Perspective on his Life and Works*,

> The Chaldeans long ago believed that the planets revolve in nearly concentric orbits around the sun and that comets do so in extremely eccentric orbits, and the Pythagoreans introduced this philosophy into Greece. But it was also known to the ancients that the moon is heavy toward the earth, and that the stars are heavy toward one another, and that all bodies in a vacuum fall to the earth with equal velocity and thus are heavy in proportion to the quantity of matter in each of them. Because lack of demonstration, this philosophy fell into disuse, and I did not invent it but have only tried to use the force of demonstrations to revive it.[14]

Further more, does not the first commandment of the original religion instruct human beings to worship and love God alone, thus telling human beings explicitly never to put any material mediator (like a sensible object) between themselves and God?[15] Is not this the prescription that corrects the distortion wrought in the sensorium? Moreover, is not "absolute, true and mathematical time," and the mathematical analysis of the motion of bodies around central forces a modern scientific return to the true ancient worship of God around sacrificial fires? Indeed, Newton writes:

> It was the most ancient opinion of those who applied themselves to Philosophy, that the fixed stars stood immovable in the highest parts of the world, that under them the planets revolved about the sun, that the earth as one of the planets, described an annual course about the sun, while by a diurnal motion it turned on its axis, and the sun remained at rest in the center of the universe. This was the philosophy taught by old Philolaus, Aristarchus of Samos, Plato in his riper years, the whole sect of Pythagoreans, and that wisest king of Romans, Noma Pompilius. As a symbol of the round orb with the solar fire in the center, Numa erected a round temple in honor of Vesta, and ordained a perpetual fire to be kept in the middle of it.... And in the Vestal ceremonies we can recognize the spirit of Egyptians who concealed the mysteries that were above the capacity of the common herd under the veil of religious rites

---

(Oxford, 1988), pp.185-201; Guicciardini, *Reading the **Principia***; Mcguire and Rattansi, "Newton and the Pipes of Pan," p. 118; De Smet and Verelst, "Newton's Scholium Generale;" Dobbs, *Janus Faces*, pp. 197-209; Snobelen, "To Discorse of God."

[14] "Unpublished Preface to the *Principia*," *Isaac Newton: The Principia*, p. 53, (from ULC MS Add. 3968, fol. 109).

[15] See e.g. Exodus, XX, 2-7.

and hieroglyphic symbols.[16]

Thus following the "purity of mathematical and philosophical truths" of the *Principia* restores a truth similar to the ancient pristine worship, which also restored the direct connection with God while marveling at the mathematical beauty and harmony of creation, especially of the solar system. From the *Principia* it is clear that Newton thought that the mathematical principles he was able to uncover govern the solar system and establish the dominion of God over creation, showing us our duty towards him. The *Principia* proves, so Newton tells us in the General Scholium, that "[t]his most beautiful system of the sun, planets, and comets, could only proceed from the counsel and dominion of an intelligent and powerful being." This truth does not remain in the abstract but has immediate practical implications regarding our duty towards God:

> We know him only by his most wise and excellent contrivances of things, and final causes; we admire him for his perfections; but we reverence and adore him on account of his dominion: for we adore him as his servants; and a god without dominion, providence, and final causes, is nothing else but Fate and Nature.[17]

Reading the *Principia* and *Opticks* as a modern form of the original religious worship helps to explain Newton's insistence on associating the absolutes with God's sensorium. Uncovering the intricate mechanism of the divine sensorium enables human beings to connect directly with God without being misled by their vulgar tendencies to bond with sensible objects. In *De Gravitatione* Newton writes regarding God's omnipresence:

> God is everywhere, created minds are somewhere, and body is in the space that it occupies; and whatever is neither everywhere nor anywhere does not exist. And hence it follows that space is an effect arising from the first existence of being, because when any being is postulated, space is postulated.[18]

---

[16] *Add MS* 3990, f. 1, a passage from the early Book II of the *Principia* taken directly from "Origins of Gentile Theology," also cited in Westfall, *Never at Rest*, p. 434.

[17] Newton, *Principia*, General Scholium, pp. 440, 442.

[18] Isaac Newton, "De Gravitatione et Aequipondio Fluidorum," A.R. Hall and M.B. Hall (eds.), *Unpublished Scientific Papers of Isaac Newton*, (Cambridge, 1962), pp. 136-7. Dobbs argues convincingly that the paper should be dated around 1684, see: Dobbs, *Janus Faces*, pp. 139-46.

Does this mean that God's existence *per se* constitutes space and time? In the General Scholium of the *Principia* Newton is clearer on this issue, saying explicitly that God is omnipresent in the world: "He endures for ever, and *is everywhere present*; and by *existing always and everywhere*, he constitutes duration and space."[19] David Gregory also recorded on 21 December 1705 that for Newton, "as we are sensible of Objects when their Images are brought home within the brain, so God must be sensible of everything, being intimately present with every thing."[20] J.E. McGuire remarks that Newton's "God is everywhere and always: his where and when are categorically different from finite things," thus space and time for him are "physical manifestations of God's existence; neither however are properties or qualities of God." Instead "space is the *Makom* of God wherein he substantially dwells."[21] Yet, if God is everywhere present, how is it that nothing affects him? How can the Supreme Being remain uninfluenced whilst "in him are all things contained and move," and corporeal things definitely change through time?[22] How can his sensorium remain uninfluenced by the physical interactions taking place within it? And how can mortals engage directly with his supreme presence?

This issue is very subtle. Newton believed that God's sensorium is absolute, true, mathematical, and pure. As such, God's absolute space can always be present, yet remain absolutely detached from the things present to itself. In Newton's terms: "God suffers nothing from the motion of bodies; [and] bodies find no resistance from the omnipresence of God."[23] But what exactly does this pure detachment mean? How can God, who is "immaterial, non-corporeal, yet all-pervasive,"[24] be ever-present to corporeal bodies through an absolute sensorium (absolute space and time)? And how can He intervene in the affairs of the world from such a detached absolute position without being influenced and becoming degraded or influenced by matter?[25] Can men follow his ways and not get caught up in the material maze? To

---

[19] *Principia*, p. 441, italics original.

[20] David Gregory, *David Gregory, Isaac Newton and their Circle*, W.G. Hiscock (ed.), (Oxford, 1937), p. 130.

[21] J.E. McGuire, "Force, Active Principles, and Newton's Invisible Realm," *Ambix*, vol. 25 (1968), pp. 154-208, 200, 201. The term *Makom* is a jewish concept of space. On this issue, see: B.P. Copenhaver, "Jewish Theologies of Space in the Scientific Revolution: Henry More, Joseph Raphson, Isaac Newton and their Predecessors," *Annals of Science*, vol. 37 (1980), pp. 489-548.

[22] For a stimulating discussion on these questions, see: J.E. McGuire, "The Fate of the Date," pp. 279-293. For a different perspective from the ones discussed here, see: A. Rupert Hall, "Newton and the Absolutes: Sources," pp. 261-287.

[23] *Principia*, p. 441.

[24] Dobbs, "Newton's Alchemy," p. 71.

[25] On this issue see: Force, "Newton's God of Dominion", pp. 88-9 and McGuire, "The Fate of the Date," pp. 279-293.

answer these questions, let us look more carefully at Newton's understanding of the nature of God's dominion over his creation.[26] To reiterate, as human beings, Newton says,

> we know [God] only by his most wise and excellent contrivance of things and final causes; we admire him for his perfections; but we reverence and adore him on account of his dominion; for we adore him as servants.[27]

The dominion of God over his servants is analogous to the dominion of our soul over our body.[28] Indeed, writes Newton, "God may appear to our innermost consciousness to have created the world by the sole act of his will, just as we move our bodies by an act of will alone."[29] The same idea is expressed in the queries of the *Opticks*: God, says Newton, is a powerful ever-living Agent "who being in all Places, is more able by his Will to move the Bodies within his boundless uniform Sensorium, and thereby to form and reform the Parts of the Universe, than we are by our Will to move the Parts of our own Bodies."[30] Newton also calls space a sensorium when he refers to the essential role of God's will and the work of active principles:

> Life and will are active principles by which we move our bodies & thence arise other laws of motion not yet known to us.... If there be a universal life, & all space be the **sensorium** of a immaterial living, thinking, being, who by immediate presence perceives things in it as that wch thinks in us perceives their pictures in the brain and whose Ideas work more powerfully upon matter than the Imagination of a mother works upon an embrio, or that of a man upon his body for promoting health or sickness, the laws of motion arising from life or will may be of a universal extent.[31]

---

[26] A thorough discussion on the scholarly research done on the scientific and theological implication of God's dominion, is given in: Snobelen, "To Discorse of God."

[27] *Principia,* pp. 441-2.

[28] *Principia,* p. 440. On the relationship of soul and body, see also: Rob Iliffe, "'That Puzzling Problem': Isaac Newton and the Political Physiology of Self," *Medical History*, vol. 39 (1995), pp. 433-458.

[29] Newton, "De Gravitatione," p. 141.

[30] Newton, *Opticks*, p. 403.

[31] U.L.C. Add. 3790, fol. 252v, cited in McGuire, "Newton's Invisible Realm," p. 205. My emphasis.

In a research on the physiology of the self (relying upon this text and similar ones) Iliffe argues that for Newton the "fact of free will and the capacity of self-motion were always held" to be evidence that the "normal laws of motion had application over a limited domain" and that there "was more to comprehending the world than was demonstrated in the *Principia*."[32] In Newton's words:

> Matter is a passive principle & cannot move itself. It continues in its state of moving or resting unless disturbed.... These are passive laws & to affirm that there is no other is to speak against experience. For we find in ourselves a power of moving our bodies by our thought. Life and Will (thinking) are active Principles by wch we move our bodies, & thence arise other laws of motion unknown to us.[33]

Over the years Newton changed his mind regarding the laws of active principles, working through such forces as generation, fermentation, and gravity.[34] According to McGuire, he considered these active principles as "intimately connected with the causation of Divine agency."[35] The lowest degree of reality and perfection belonged to matter and its passive laws (*vis insita*) and as such they were completely dependent on the will of God. Yet "providence planned the world so that divine power should not be limited, and so that it would be manifest to those who studied Nature with piety. From this point of view active principles are the expression of God's *potentia absoluta* in Nature, such that laws and forces in the present natural order are used intermittently to actualize the pre-conceived ends of Providence."[36] Though passive, matter is not excluded from this spiritual divine plan. Instead, Newton's conception of matter recalls the ancient doctrine regarding a chain of being stretching from the throne of God into Nature. This doctrine assumes that "existing entities move imperceptibly up the chain from material through immaterial to the greatest reaches of spirituality."[37]

Matter as such, though passive, necessarily contains, hidden within it, a seat for spirituality, so to speak, otherwise how could God, the highest on

---

[32] Iliffe, "Isaac Newton and the Political Physiology of Self," pp. 453, 457.

[33] U.L.C. Add. 3790, fol. 619r, cited in McGuire, "Newton's Invisible Realm," p. 171.

[34] On this issue see: McGuire, "Newton's Invisible Realm;" Dobbs, *Janus Faces*; Iliffe, "Isaac Newton and the Political Physiology of Self."

[35] See e.g. discussion on next section on gravity.

[36] McGuire, "Newton's Invisible Realm," pp. 206-7.

[37] McGuire, "Newton's Invisible Realm," p. 186. In "The Fate of the Date," McGuire even shows that in the *Principia* Newton's God "is the ground of all being – the spiritual *tonos* and "structuring structure" of the cosmos," p. 295.

high on the spiritual scale, create, sustain and reform this most degraded entity? It is my opinion, that this dormant spiritual seat can be found in the *vis insita* of matter, once we grasp the role of mathematics and its spiritual features in the Newtonian system. Newton says explicitly that purity exists in mathematics and that the whole program of the *Principia* was to uncover the mathematical principles governing the motion of physical bodies.[38] Purity is a positive quality associated with spirituality. It is clear from Newton's writings that he thinks the purity of mathematics will help human beings to reach God's absolutes. As such, mathematics also has a religious and spiritual function. It is the language, which enables humans to narrow the gap between God and matter by understanding the general laws of nature through which God governs and sustains creation. Mathematical principles are natural laws that shape and restrict the possible actions and reactions of bodies to external forces. Essential in this context is the newly defined physical-mathematical entity which Newton calls inertial mass.

Newton conceived the inertia of mass, that is the *vis insita* of matter, to be a spiritual mathematical mechanism instilled within matter to assist God in the operation of creation. I interpret Newton's *vis insita* as a spiritual mathematical seed planted deep within matter, a kind of reminiscence of the equable mathematical flow of God's absolute time. This passive force was planted there so that Providence could sustain the creation on a daily basis and intervene periodically when necessary without being influenced by matter.[39] If God had not instilled such a passive force in matter He could not have influenced it without being influenced in return. This point supports my previous argument that Newton's method of fluxions contains his theological beliefs and gives some insight into the scientific meaning of the worship around the sacrificial fire.

I argued in Chapter II that Newton's whole project was to recover the hidden design of God's works; it was therefore crucial for him to understand the internal hidden mechanism of material bodies. In his search he discovered a sophisticated mechanism of a passive force instilled in material atoms, which is governed by a mathematical lawfulness defined by the method of fluxions. *Vis insita* is never observed in nature yet its mathematical behavior can be detected through a body's resistance to any change of its physical state. Why did God design matter with such a protective yet aggressive passive force that obeys mathematical laws and is intrinsically connected with the mathematical nature of the absolutes? This innate passive force is the mechanism through which corruption and loss of motion enter the divine design. At the same time divine order is constantly and effortlessly preserved and sustained in matter. Here I will present my conclusions as simply as possible adducing the evidence in later sections. There I will suggest that we have good reason to believe that these were Newton's concepts, yet knowing how heretical these ideas were at the time, he left the work of interpretation to later readers, as the ancient

---

[38] Newton, *Principia*, pp. 18, 319.

[39] A similar view is given in: McGuire, "The Fate of the Date," pp. 291-2.

prophets left the work of interpretation to scholars like himself. The deeper the divine truth the more hidden it should remain less it be misused.[40] This subject will be further discussed below.

Let us first look at the philosophical implications of the notion of *vis insita*. As long as a body is not interrupted by any external force it remains in its inertial flow. This inertial flow is pure in the sense that it flows equably along God's absolute time in a mathematical relation to absolute space and time and can be calculated through the purity of a mathematical analysis. When an external interaction disrupts the body's equable flow the passive force is programmed to protect the inertial state of the body and repel in return according to a mathematical proportion that reacts lawfully towards the interruption. The whole analysis of the reaction can be detected in the mathematical relationship it has to the absolutes at the nascent moment of the interruption. It is in this hidden passive material design that I find the analogy between idolatry and true religion. Idolatry is a form of worship that is no longer connected with the true divine source and as such has stopped being as lawful as it was designed to be. God's lawful design is simple and pure as the equable flow of time. Only when momentary disruptions occur does the simple flow of atoms become complex. Similarly, idolatrous actions introduce corruption into the original design of humanity because at certain moments in history human beings stop obeying the first commandment of worshipping God alone. But what is the meaning of becoming disobedient to God? Mathematics is the best language to describe the absolute lawfulness God demands from human beings and the consequences of disobedience to the manual of the design (the commandments). It is the purest, the most restrictive and most law-abiding language existing; therefore, it is the fittest of all languages to describe God's governance over creation. In physics the language of mathematics is able to uncover the law-abiding regularity God has imposed upon matter through his mathematical sensorium, as well as the consequence of complexity arising from disruptive interactions. In human history the hidden regularity is more complicated but a similar mechanism is present. Newton criticizes the vulgar notions of physics which people construct in relation to sensible objects. True notions in physics are mathematical and defined in relation to God's absolutes alone. Another analogy is implicit in Newton's system, this time between the passive inertial force of bodies and the will of human beings.

Newton's bodies are designed with an internal mechanism similar to the will of human beings. As long as the human will follows the divine Will it obeys God's commandments completely. Idolatry is a form of human will following its own dictates. Within this structural analogy, bodies can also worship God properly with the least effort when they flow along God's equable flow of time, or they can become idolatrous by interacting violently in material interactions that cause a loss of motion and a loss of order to the

---

[40] On the exoteric and esoteric strata of Newton's writings, see: Snobelen, "God of gods," pp. 204-208; Faur, "Newton, Maimonides, and Esoteric Knowledge."

whole worldly design. The design is so wise yet so simple that as long as bodies are connected with God's absolutes and nothing else they flow equably effortlessly according to the divine state that God imprinted upon them at creation. Deviation from God's equable flow of time and a loss of the God-given original structure occurs only if there is a disruption by an external force (gravitational forces are an exception to this rule since they are active divine forces that assist God in sustaining the design).[41]

A similar design is manifested in the material passive force of bodies and in the will of human beings. In bodies the inner passive mechanism may operate according to its original design, sustaining the order God imprinted upon matter, or it may be misused or disrupted by external forces. When a disruption occurs there is a price to pay. This price (loss of motion due to a momentary change in the equable fluxion) can be calculated mathematically since the internal design of bodies is such that mass resists change to its inertial state according to mathematical laws in relation to God's absolutes. Mathematics is thus again a law-abiding language that calculates a just toll, so to speak, according to the nature of the disruption. In the human domain a similar phenomenon occurs though it cannot be calculated as accurately as in the material, since people are more complex than bodies. As long as humans are connected with God alone (and also follow the second commandment of loving the other like yourself) they enjoy prosperity, peace and happiness; but when they become idolatrous, corruption enters into society. Corruption in human affairs is analogous to material interactions that cause loss of motion. Human corruption and loss of motion are states of the world through which disruption and loss of order affect the original design. In Newton's system wrong action, that is, a mishandling of the original design, creates a loss of order. The material and human realms display a similar pattern. The difference is that matter, being the lowest entity in the scale of being, is not aware of its spiritual design. It possess, so to speak, only a dormant plan of the human and divine will that flows equably along God's absolute time. This is also why the method of fluxions can describe the behavior of bodies but not that of the human will, since bodies are less complex systems than human beings. Yet both bodies and humans pay a severe price for disobedience to the laws given by the governor, a loss of order to the worldly system occurring in both cases. But why is this equable flow so essential to the running of the design? James Force suggests that "Newton's calculus is based on the continuity of flow as supervised by the God of Dominion operating in his generally provident mode of creator and preserver of the current state of natural law."[42]

Indeed, when human beings follow God's commandments they operate in the world like God, acting without being reacted on in return. Their actions do not cause a loss of order to the original design. The state of having a direct connection with God whilst interacting in a loving manner with other human beings preserves the original design; this is the direct outcome of

---

[41] This topic is further elaborated in the next chapter on the pristine religion and alchemy.

[42] Force, "Newton's God of Dominion", p. 88.

following the true religion. Yet there is one major difference between *vis insita* and the human will. The spiritual remembrance of the divine flow inherent in matter could not by itself sustain the divine order. If this passive force, which resists any change of the inertial state of atoms, were the only spiritual force working upon matter, Newton says, "the Bodies of the Earth, Planets, Comets, Sun, and all things in them would grow cold and freeze, and become inactive Masses." Indeed, "the variety of Motion which we find in the World is always decreasing, [and] there is a necessity of conserving and recruiting it by active Principles."[43] The order of the world decreases in time since interactions and external forces draw atoms out of the original order given to them by God and they react in return, causing more disruption. Even the innocent interaction of rays of light with bodies may cause a stir and a reaction in the internal constitution of bodies:

> Nothing more is requisite for putting the Rays of Light into Fits of easy Reflexion and easy Transmission, than that they be small Bodies which by their attractive Powers, or some other Force, stir up Vibrations in what they act upon, which Vibrations being swifter than the Rays, overtake them successively, and agitate them so as by turns to increase or decrease their Velocities, and thereby put them into those Fits.[44]

Furthermore, Newton was aware that "motion is ever lost by communication especially twixt bodys of different constituents: and therefore it can no way be conveyed to ye sensorium so entirely as by the aether it selfe."[45] Thus it is reasonable to conclude that bodily senses are intermediary channels through which human beings gain information in the sensorium regarding the created world. This means that human beings' sensory organs are prone to deceive them as historical time goes on, since most people do get caught up in direct material interactions with sensible objects. The only remedy for such disruptions is to follow God's commandments. God gave humans free will so that they may choose to follow the commandments or to disobey him. If they follow the commandments of the original religion their will becomes associated with the divine will since they do as He wills. If they disobey the commandments they follow their own desires and disrupt the original order. This kind of disobedience happens very often in history and when human non-compliance to God's commandments disrupts the order to a point of no return, God must intervene in worldly affairs in order to repair the damage. God's periodical intervention is a reorganization or renovation of a worldly design

---

[43] *Opticks*, pp. 399-400.

[44] *Opticks*, pp. 372-3.

[45] Newton, J.E. McGuire and M.Tamny (eds.), *Certain philosophical questions: Newton's Trinity notebook*, (Cambridge, 1983), p. 488.

that has lost its original order due to a misuse of the manual.

God's will governs all aspects of creation.[46] As long as human beings follow God's Will (the commandments) their will becomes divine in the sense that they choose to obey and worship him alone. Their personal will becomes a sort of divine conduit. By following God's commandments people operate in a divine manner, enjoying the design at its best. Disobeying God means that the human will does not choose to connect with God alone and begins to indulge and be caught up in material bondage (through worshiping stars and other material objects such as idols, or other human beings). The beauty of the design is that God has endowed matter with a mechanism similar to the human will, called the *vis insita*, which enables God to preserve and sustain any divine order he chooses with the smallest expenditure of effort. As long as the creation is lawful according to God's will (the original order of the design) no disruption occurs within the daily operation since the *vis insita* of matter preserves whatever state God has imprinted upon it. Yet when human beings start to disobey God's will the whole worldly order gradually becomes corrupted since the human participants of the system are no longer as restricted and lawful as they were originally designed to be. Disorder, complexity and chaos enter the system, also indirectly influencing the material ordered design.

II. The Role of Gravity

As Dobbs says, Newton searched for the cause of gravity throughout his life.[47] His exposition of the nature of gravity underwent startling and dramatic changes as he examined certain questions and discovered relevant scientific evidence. In the first period of his life he searched for agents, acting as intermediaries, between God and the world that could explain the variety of living forms as well as the cause of gravity and heaviness. In those early years, long before the *Principia*, he held that the alchemical agent could be taken as the intermediate agent that assists God in the continuing governance of the world. He also associated Christ with the role of the alchemical agent.

---

[46] James Force sums up God's masterful governance succinctly as follows: "Newton's view of God's Dominion, i.e., the total supremacy of God's power and will over every aspect of creation, colors every aspect of his views about how matter (and the laws regulating the ordinary operation of matter) is created, preserved, reformed, and occasionally, interdicted by a voluntary and direct act of God's sovereign will and power. Newton's commitment to the Lord God of Dominion issue necessarily the dependence of Nature upon God's will. He creates it and (at the same time) he creates it to operate by the ordinary concourse of the laws of Nature. He preserves it, he reforms it, from time to time he directly suspends its ordinary operation through a specifically provident act of will, and He has promised in prophecy to destroy it as the wise have good reason to understand and to believe it." Force, "Newton's God of Dominion," p. 84.

[47] Dobbs, *Janus Faces.*

Combining his alchemical studies with his Arian beliefs he assigned Christ the part of Logos, also proposing that Hermes Trismegistus, the founder of alchemy, was an earlier form of Christ.[48] Newton's alchemical and theological manuscripts of this period show that he believed the existence of this intermediate agent to be a certifiable evidence of providence. He also considered the possibility that the body of light permeating everything might be the agent for the alchemical vegetable spirit. In the alchemical tradition this spirit was considered the agent of alchemy since illumination was sometimes associated with the light of genesis and it represented the power to activate or reactivate lifeless matter.[49] Turning to the cause of gravity, we find that in the 1660s Newton considered gravity to be a mechanical mode of action. In the 1670s he assumed that "the more 'spiritual' an agent is, the finer the matter of which it is composed, or, conversely, the smaller the material particle, the greater its spirituality and activity."[50] During that period, he partially fused mechanical gravitational aether with non-mechanical alchemical spirit.[51]

However, when Robert Hooke and Edmond Halley challenged him to undertake the *Principia,* Newton abandoned his former ideas of a subtle corporeal active gravitational aether. In the next three decades he gave preference to a spiritual cause of gravity, for he concluded that only a spiritual cause could penetrate adequately into celestial bodies without consisting a frictional drag on the motions of the planets and comets and producing a slowing down of their motions, contrary to observed phenomena.[52]

Dobbs points out that during the first period after the *Principia* (1687-1713) Newton shifted his focus from alchemy to a study of natural philosophy, since he believed this would be the best way to restore the true religion and give a satisfying answer to the question of gravity.[53] The ancient texts suggested two solutions to him: either God subsumed gravity directly or an intermediate agent existed which did not constitute a drag on the motion of the heavenly bodies.[54] At this period the ancient texts, especially the Jewish idea of God as *makom* (place), provided a source for his idea that the omnipresence of God constitutes absolute space and time, and they also offered him a solution as to the nature of gravity. Newton himself defined the problem as follows:

> By what means do bodies act on one another at a distance? The ancient philosophers who held Atoms and Vacuum attributed gravity to atoms without telling us the means unless in figures:

---

[48] Dobbs, *Janus Faces*, p. 37.

[49] Dobbs, *Janus Faces*, pp. 39-40.

[50] Dobbs, *Janus Faces*, p. 96.

[51] Dobbs, *Janus Faces*, p. 98.

[52] Dobbs, *Janus Faces*, p. 191.

[53] Dobbs, *Janus Faces*, p. 170.

[54] Dobbs, *Janus Faces*, p. 191.

> as by calling God harmony representing him and matter by the God Pan and his Pipes, or by calling the Sun the prison of Jupiter because he keeps the planets in their Orbs. Whence it seems to have been an ancient opinion that matter depends upon a Deity for its laws of motion as well as for its existence.[55]

Though the ancients did not explain exactly how the Deity put the laws of motion into effect, Newton derived a more precise idea from the Stoics:

> ... [T]hose ancients who more rightly held unimpaired the mystical philosophy as Thales and the Stoics, taught that a certain infinite spirit pervades all space *into infinity*, and contains and vivifies the entire world. And this spirit was their supreme divinity, according to the Poet cited by the Apostle. In him we live and move and have our being.[56]

Newton saw his thought on gravity "not as something new under the sun but rather as a step toward the restoration of the true natural philosophy of the ancients."[57] Indeed, Rudolf De Smet and Karin Verelst have pointed out a number of verbal and conceptual parallels of Newton's definition of God's omnipresence, oneness, and unity (as they appear in the General Scholium) and the works of Philo Judaeus (the first-century Jewish philosopher).[58] God's omnipresence and his universal gravity offered Newton an unexpected opportunity to demonstrate direct divine activity in the created world. It was founded on the arbitrary will of God. These concepts point out that at this period (including the period of the General Scholium) Newton's God had a "thoroughly Hebraic and Biblical character."[59] Newton was cautious in speculating on gravity in public, yet in private he presented his theological speculations much more openly.[60] For example, the Scottish mathematician

---

[55] Isaac Newton, draft Query 23, Portsmouth collection MSS Add. 3970, f.619r, as quoted in Mcguire and Rattansi, "Newton and the Pipes of Pan," *Notes and Records of the Royal Society of London*, vol. 21 (1966), pp. 108-143, p. 118.

[56] Newton, Portsmouth collection MS 3965.12, f.269, cited in Mcguire and Rattansi, "The Pipes of Pan," p. 120.

[57] Dobbs, *The Janus Faces*, p. 210.

[58] De Smet and Verelst, "Newton's Scholium Generale."

[59] Snobelen, "God of gods," p. 177.

[60] On Newton's speculations about the divine cause of gravity, see also: John Henry, "'Pray not ascribe that Notion to Me': God and Newton's Gravity," *The Books of Nature and Scripture*, pp. 123-147; Michael Heyd, "*Be Sober and Reasonable*:" *The Critique of Enthusiasm in the Seventeenth and Early Eighteenth Centuries*, (Leiden, 1995), pp. 244-251.

David Gregory wrote on Newton at the time: "what cause did the ancients assign of Gravity? He [Newton] believes that they reckoned God the Cause of it, nothing els, that is no body being the cause; since every body is heavy."[61] Newton also speculated in privy with Nicolas Fatio de Duillier, Christopher Wren and William Whiston on God and gravity.[62] Yet much work is still needed to elucidate what precisely Newton had in mind when he said that the will of the supreme and omnipresent Deity operates gravity directly. How can divine will operate gravity directly? And what precisely is gravity in such a theological context?[63]

From 1713-27, in the last period of his life, Newton searched for a less spiritual and abstract explanation for gravity because he was dissatisfied with his in-conclusive spiritual solution. Newton's final version of the gravitational aether stemmed from his exposure to electrical experiments. From these experiments he concluded that a connection exists between electricity and light and that the electrical effluvium is a source of activity in micro-matter working with light to stir up and organize the particles of passive matter in living forms. He extrapolated from the elasticity of the electric spirit a newer version of an exceedingly elastic and active aether intermediary between the incorporeal nature of God and the full corporeality of body. This aether explained the work of gravity. Dobbs argues that behind the search for such an intermediary substance was also the theological drive stemming from Newton's lifelong Arianism. The Arian beliefs suited the new version of active aether since God could again have an agent, a viceroy, through whom he could create and govern the world. This aether was a "mediator that brings divine ideas into the world and insures that they are embodied in their variety. This Mediator is also in charge of gravity and renovation and functions similar to Christ and the Word. It is an Agent who put God's will into effect in creation and from then on."[64]

In what follows I shall try to comment on Newton's solution in the first period after the *Principia* (1687-1713). Newton's writings of this period should be situated in the context of his historical and theological writings on the ancient religion and his concept of the Hebraic God of dominion.[65] His

---

[61] Gregory, *David Gregory*, p. 30.

[62] *The Correspondence*, vol. 3, pp. 308-9; vol. 4, pp. 266, 267; Whiston, *A Collection of Authentick Records belonging to the Old and New Testament* (London, 1727), vol. 2, pp. 1072-3. For a fuller discussion, see: Snobelen, "To Discorse of God."

[63] These inquiries are further developed in the section on Newton's Cosmology.

[64] Dobbs, *The Janus Faces*, pp. 247-8.

[65] See: Snobelen, "To Discorse of God." Snobelen points out that Newton's God of dominion goes hand-in-hand with his antitrinitarianism. On Newton's antitrinitarianism see: Dobbs, *Janus Faces*, pp. 213-49; Force, "Newton's God of Dominion," pp. 75-102; Larry Stewart, "Seeing through the Scholium: Religion and Reading Newton in the Eighteenth Century," *History of Science*, vol. 34 (1996), pp. 123-65. Newton's Arianism is related to the concept of the God of dominion but is not as clear as Dobbs argues. On this see Mcguire's work on the tension in Newton's writings between the

radical proposal regarding God's Will as operating directly upon matter as gravitational force is baffling. How can spiritual will influence passive matter directly through an active force called gravity? What is the divine will? Does it share any kind of constitution with gravity? What is the relation between divine will, gravity and matter? How can differing substances operate directly one upon the other? To approach these delicate inquiries, I shall attempt to present a philosophical model taken from computer engineering. This model, I trust, will assist elucidating in the spiritual meaning Newton may have assigned to God's dominion, omnipresence, and gravity. The model connects the three laws of motion including gravity and the two central commandments of the ancient religion. My interpretative model may contribute some elements to Newton's writings of the first period after the *Principia*, of coherence and meaning though of course he himself could not have developed such a basis of argumentation.[66] The model may also give a certain insight into the connection he makes between God's will, gravity, and material atoms, setting out his notion of the bond between the religious commandments of the ancient religion and the laws of nature, and of how both sets of laws fitted together in the providential plan. It may also explicate the two different roles of time (eternal absolute time and the historical time of the created world), the analogous structure of *vis insita* and human and divine will, and the deeper meaning of the created world as a divine design.

III. God's Worldly Design as Computer Design

In computer science the idea of an original material design suddenly going wrong due to human error is easy to appreciate. Computer hardware functions rather like Newton's definition of the laws of nature, whereas the software can be compared to the law of gravity, and the specific computer programs are what Newton called the original design God imprinted upon matter. The electrical current that is essential for the operating of the computer functions like God's daily sustenance. The hardware is the basic framework of the computer, composed of electrical and logical circuits into which engineers insert computer software that instruct the machine to perform mathematical and logical operations. The simplest computer language is the binary one, of a

---

Arian God who "is transcendent and works in nature though an intermediary" and "the God of dominion of the 'classical scholia' who is directly present and active in creation." McGuire, "The fate of the date," p. 294. On the Hebraic character of the God of dominion in the General Scholium, see: Manuel, *The Religion of Isaac Newton*, pp. 16-7, 20-2, 40, 74-6; Snobelen, "God of gods,"; De Smet and Verelst, "Newton's Scholium Generale."

[66] I am well aware that this model is non-historical and could have not been available to Newton. I considered dismissing it, yet throughout my teaching and lectures on the subject the audience found the model helpful since it presents abstract spiritual ideas in simplistic terms familiar to all.

structure similar to the currents of the hardware, because in electrical circuits you either have a current or you don't. From simple binary language more complex languages can emerge to handle more complex situations. The computer can be operated as a tool for specific tasks only according to the defined instructions of the programmer. Any computer program can be seen as a human design inserted into matter. On this analogy, the hardware infrastructure and the software function like the few and simple constant laws of nature upon with which one can compose many programs. Similarly, Newton's three laws of motion define the infrastructure of all material bodies (that is, the infrastructure of inertial and gravitational mass) into which God inserted many multi-leveled designed orders (such as the solar system, the human body, etc.). Indeed each physical system (which is an abstraction of a small part of the worldly design) described in the *Principia* is governed by the laws of motion in the way each computer language is conditioned by the infrastructure of the laws governing the electrical circuits of the computer.

All the operating systems in the computer are conditioned by the infrastructure of the hardware and software, as Newton's mathematical principles of the system of the world govern each material aspect of creation. The three laws of motion describe the infrastructure of passive matter, that is, its resistance to any change of its inertial state, and the mathematical lawfulness in which it acts and reacts when the inertia is disturbed. Similarly the infrastructure of the computer is such that it is conditioned by the laws of electricity defining how an electrical current will act, in any given system. Before engineers knew these electro-magnetic laws of nature they could not develop any kind of appliance similar to the computer. Knowledge of the laws of nature enabled them to design many operational systems governed and conditioned by these laws. Newton himself refers to his mathematical principles as "laws and conditions of certain motions, and powers or forces," and he shows how "from the same principles," he will "demonstrate the frame of the System of the World."[67]

The frame of the system of the world is the multi-layered original design God chose to insert into passive matter, as any given computer program is inserted into the hardware and software of a computer. Following out this analogy the design of the world system appears to be a matter of divine choice, because it is possible to insert multifarious designed informative programs into the simple laws of nature, as it is possible to change the programs of any computer according to the wishes of the user. The worldly design is like a huge computer with many operational systems working in equivalence, all conditioned by the same simple infrastructure.

To operate any given program all one needs is to learn and follow the manual of each of its components. The user does not have to understand the infrastructure and the software. All that is necessary is that the manual should be written in a comprehensible language. This language has nothing to do with the computer languages according to which the design was written. Though

---

[67] Newton, *Principia*, p. 319.

everything is well defined in the operational system of the program we all know that there are many ways in which the operation can go wrong if one does not strictly follow the manual. Further, it is clear that a human mistake in one component may cause damage to another since everything in the computer is inter-connected. There is one way to operate the system correctly and many ways to get it to go wrong by the insertion of mistaken instructions that disrupt the flow of the program. When we speak about a multi-leveled design inserted into matter it is obvious that a wrong move can disrupt or even cause parts of the material design to collapse.

The computer analogy is also helpful in understanding why Newton's God needs to sustain the world on a daily basis. To operate any software computer program the computer has to be plugged in. As uninterrupted electrical current is necessary for the operation of the computer, God must uninterruptedly sustain the world. The program needs a supply of electricity without this it remains dormant. But whence comes the flow of divine energy? It seems to me that Newton's absolutes and the method of fluxions provide an answer. The Newtonian God's omnipresence in space and his absolute equable flowing time are the mechanism through which he sustains creation. Absolute space assures that all designed bodies will have a direct contact with the supreme supplier of energy. God's space is like a huge reservoir of divine energy through which he daily sustains the world. The equable flow of time, in turn, is the constant unconditional energy (the current) that flows from God's will directly through space to all materially designed bodies through their *vis insita*.[68] The *vis insita* is the necessary piece of equipment instilled into matter for transferring the free and unconditional divine flux called absolute mathematical time without which the whole designed world system would remain static.

The computer works on similar principles since the electron is the particle necessary for the occurrence of the electrical current. To utilize the current engineers need wires and an energy source. Similarly, absolute space and the material bodies that God has created are the medium through which he can transfer the free energy of equable flow that runs the worldly design. Matter is designed within its infrastructure with a passive inertial force, assuring that the material world (and not only the human) will be attuned and plugged into the unconditional divine energy. For this reason the inertial state of a body is retrieved after each momentary disruption even though the original order was damaged in some way. *Vis insita* is a design made specifically in order that any material atom can be effortlessly sustained by the divine absolute equable flux (called absolute time); thus unless the body's inertial state is disrupted it will continue to be charged by the equable flux.

Actually, two mechanisms are available for daily sustenance, the

---

[68] This idea is compatible to the modern theory in physics pertaining to the latent energy of the vacuum. In fact it is believed by several researchers that our universe as well as parallel universes were created as the result of the fluctuation of the energetic vacuum. See: A. Linde, "The Self Reproducing Inflationary Universe," *Scientific American*, vol. 271, no. 5 (1994).

inertial mass and the gravitational mass. The inertial mass is the direct relationship of each particle of matter with the divine space and equable flow of time, whereas the gravitational mass is the mechanism through which the cables of the design carrying God's flowing current are inserted into the infrastructure. Inertial mass is the mechanism corresponding to the first commandment of worshipping God alone, and gravitational mass is the mechanism underlying the second commandment, the loving relationship among the members of the designed system. Gravity is the relation of each massive body to all other bodies and it is also a mutual relationship that obeys mathematical lawfulness. It is inertia and gravity that enable God to sustain the design with the least output of effort. All God needs to do to run the program is to radiate his eternal equable flow (which he does anyway). This radiation automatically defines both the inertial mass of the body and its gravitational output. In other words, the *vis insita* of a body is sustained constantly through the equable flux of divine time proportional to its specific material design. This constant sustenance establishes the body's inertial mass and also automatically its gravitational interactions with other massive bodies, since the amount of inertia that a body can capture and sustain from the divine source is what the body spreads automatically as gravitational forces to all other bodies within the worldly design. The more massive a body is the more equable flux enters through it to the worldly design, spreading automatically as a gravitational current in proportion to the distance and the material constitution of other bodies that undergo the same process. The following Newtonian text conveys this idea of gravity being God's eternal flux spread proportionally according to the distribution of mass in space:

> For two planets separated from each other by a long distance that is empty do not attract each other by any force of gravity or act on each other in any way but by the mediation of some active principle interceding between them by which the force is transmitted from one to the other. And therefore those Ancients who rightly understood the mystical philosophy taught that a certain infinite spirit pervades all space & contains and vivifies the universal world;... By this Symbol the Philosophers taught that matter is moved in that infinite spirit and is acted upon it, not in an irregular way, but harmonically or according to the harmonic ratios as I have just explained.[69]

Put differently, the daily sustenance is the Godly equable current (fluxion), which is the energy needed for the running of the whole design. Human beings, who are part of the design, call its running the historical time of the world. God's equable flow (his eternal duration) is the constant current that makes historically designed world time run. Without this constant current

[69] Newton, Add MS 3965.6, f. 269, cited in, Westfall, *Never at Rest*, pp. 511-2.

entering the worldly design no historical time would be composed. The two different times, God's equable flow and historical time are analogous to an eternal current that keeps the historical time of the program running. The worldly program can be turned off or on according to God's choice. Such turning on or off never touches God's constant current (equable absolute flow). Similarly, in a computer the current that runs the program is constant, yet the program has its own historical time that changes from moment to moment as long as we do not disconnect the computer from the electrical source.

This computer analogy illustrates how and why a human error in applying instructions can disrupt the design immediately and also why a small insignificant mistake expands as time passes. However, we need to differentiate between the hardware and software of the program and the order of the world system. The software (gravity) is expressed in the mathematical description of the lawfulness of the universe, which in turn forms the order of the design depending on the initial conditions God chose in creation. God chose initial conditions that created a beautiful, symmetrical, harmonious and stable order. As time passed the equilibrium of the system kept being disturbed owing to a partial misuse of the manual. God gave the material design a sort of freedom allowing it to choose, so to speak, whichever order it wants depending on the initial conditions. Once the equilibrium of the worldly order is disturbed, depending on the severity of the disturbance, the material design is such that it can become receptive to the insignificant initial conditions of the disturbance and consequently change its divine order dramatically as historical time passes. However, the mathematical tool developed in the *Principia* describing the lawfulness of the world, cannot describe the outcome of complex systems beyond a certain threshold. Indeed, beyond this threshold, the mathematical formulas which Newton discovered describing the original divine order imprinted upon matter by God at creation, are no longer of use since the simple order has become too complex to compute. Only God has the ability to comprehend all the aspects of this complexity but he has bestowed only part of his wisdom upon humans.

No mysterious mechanism is needed to explain how a human error can influence the material design. Matter is conditioned both in its infrastructure (by the laws of motion and gravity) and the chosen order to follow God's will, whilst his equable current (equable fluxion) runs the program. As long as everything remains lawful according to the Creator's intention, the flux of energy is unconditional. When an unlawful action has occurred the current entering the system becomes less sufficient since at a certain point there is a blockage due to the insertion of wrong instructions. At first the blockage may be small but if the same mistake keeps recurring the blockage grows and the mistake becomes a habit since the change has been inserted also in the infrastructure. The third law of action and reaction ensures that for any activated human mistake there is a toll, since every action is inserted into the software of the material design. There is no action in the worldly design (including the actions of human beings) without a reciprocal reaction from the material ordered design. From the perspective of the worldly design each human being is a very sophisticated local program, and the specific program of each human being runs equivalently to the orders of other human beings. If a few of these local programs become idolatrous the whole order of the design may collapse as historical time passes, even if there is no direct connection between the "electrical and logical circuits," so to speak, of the human and material orders. The worldly design is like a huge computer program sustained by the absolute current of God. The divine current (the daily sustenance) will keep the whole program running even though a few local programs have created blockages due to the insertion of wrong instructions. Yet in the long run each mistake becomes more and more destructive since it keeps dispersing the free current, thus automatically and unintentionally causing changes in all bodies due to the automatic spreading of the current. A mistake in one local body necessarily disrupts the whole order of the design since gravity is reciprocal. Human beings play a more significant role than bodies in the worldly designed order because they are the only programs that have free will to be either lawful or disobedient. Thus their actions are significant for the running of the design although their material composition is insignificant in relation to the celestial bodies.

The two commandments of the original religion are actually simple instructions on how to operate the design with minimal effort. Connecting to God alone is the best way to capture the eternal current because only then does the human will function as a pure divine conduit assuring that no wrong instillation of unplanned circuits will be created due to material worship. Idolatry creates unplanned complex circuits within the infrastructure of the material design, as if the basic unit of the program –a human being – instead of plugging in only to the divine current were to start following another local program inserted within the infrastructure. Idolatry thus becomes disastrous to the original design because less divine current enters the system and additional disruptive unplanned complex circuits keep being inserted in the design. The second commandment of Noah's religion further explains the correct running of the program. Divine love is unconditional, as the constant divine current of absolute time is unconditional to the worldly designed program. It is only when human beings become idolatrous that this love becomes conditional since the constant current becomes less available. Why should a human being programmed in such a way that he can plug in directly to the divine current ever choose to connect (worship) directly to another material design (such as stars or other human beings) within the design. This is idolatry, the shortsightedness of human beings and their preference for material substitutes over God's pure mathematical and spiritual absolute fluxion. Material corruption is a loss of comprehension of the worldly design since it obstructs God's truth and makes what is simple and ordered in the design complex and superfluous. Idolatry is an insertion of complexity into the divine order, blocking the unconditional free divine current necessary for the running of the program. This is also the meaning of loss of motion in the solar system. God needs to intervene when his children have become idolatrous that they no longer remember him, his unconditional current hardly enters the material design and the whole program is on the verge of collapse. God then replenishes the order of the design with comets and reminds humanity of the true manual through a spiritual messenger so that his unconditional fluxion will once again run the program.

A major departure from the computer analogy is the justice and wisdom according to which the divine program is designed, in contrast to the efficiency that underlies the humanly designed computer. In Hebrew there is an idiom "מידה כנגד מידה" (*midah k'neged midah*) which carries the same meaning as the saying – "as you sow so shall you reap."[70] In Hebrew the word *midah* (value) also means quantity. This is the wisdom and morality underlying the third law of motion. In God's worldly design a just outcome responds to any operational error. The morality of "as you sow so you shall reap" is inserted passively and mathematically into the infrastructure of matter. Once a person does not follow the commandments and becomes idolatrous in one way or another he damages the divine design, in proportion to his wrong doing he will be cut off from the free divine equable unconditional current running the program. At a certain moment in history it may seem to an idolator that he is happier, yet in the long run he will lose his divine nature. A charismatic leader who does not connect his followers to God is able to create an inner program through which he takes the current of those who admire him, yet in the long run they will all be punished and removed from the divine current. Those who worship the leader or any other material substitute are disconnecting themselves from the real source of the current. After a while the energy of the whole mini-system of worship will collapse, spreading its errors to every corner of the world. This kind of justice obviously does not guide the design of a computer, since a human being is not punished proportionally to the mishandling of the manual. If the computer falls due to human error the person does not crash with it. But in the divine system the bodies of human beings are part of the worldly design. When they mishandle the design they are cut off from the unconditional equable flow which is necessary for their own well-being and the running of the world. Complexity and dis-function enter the world system when the laws are disobeyed. Complexity is a form of the disorder of an original simple, hierarchical and symmetrical order.

The analogy of the computer was given in order that my insights on the spiritual dimension of the Newtonian project will be as clear as possible to the reader. What I propose to do now is to point out many ideas from Newton's scattered writings and manuscripts that set me off to reach such an interpretation.

---

[70] See this morality, e.g. in: Exodus, XVI, 24.

IV. NEWTON'S COSMOLOGY

In order to discuss the role of passive and active forces and the total dominion of God over creation we need to turn to Newton's cosmology.[71] Newton maintained that God created a cyclical cosmogony, which He constantly sustains and replenishes when necessary. Pierre Kerszberg argues that Newton believed we perceive "the action of God wherever Nature is not simply uniform and indistinct. Any configuration would be real because of God's sovereign will."[72] Indeed, according to Iliffe, in *De Gravitatione* God's will is presented as so great that Newton can put "forward a notion of space in which there are always a multitude of figures not disclosed to sight. For Newton these objects *actually* existed, although they became visible only when God endowed them with sensible qualities (in the same way that dye made visible swirling figures in water)." Bodies "existed by Divine Will" since the power of God "was such that he could have created bodies in an infinite number of ways."[73] Thus creation is a consequence of God's wish to design and operate a certain order upon passive matter. The *vis insita* alone can not explain the motion of bodies and the non-uniform distribution of atoms in space. Indeed, in a series of letters to Richard Bentley on the nature of gravity, Newton insists that the world would remain in primeval chaos, that is, matter would be distributed in perfect uniformity in space unless God chose otherwise:

> [T]he hypothesis of Matter's being at first evenly spread through the Heavens, is in my Opinion, inconsistent with the Hypothesis of innate Gravity, without a supernatural Power to reconcile them, and therefore it infers a Deity. For if there be innate Gravity, it is impossible now for the Matter of the Earth and all the Planets and Stars to fly up from them, and become evenly spread throughout all the Heavens, without a supernatural Power, could never be heretofore without the same Power.[74]

---

[71] On this issue see: David Kubrin, "Newton and the Cyclical Cosmic: Providence and the Mechanical Philosophy," *Journal of the History of Ideas*, vol. 28 (1967), pp. 324-346; M. A. Hoskin, "Newton's Providence and the Universe of Stars," *Journal of the History of Astronomy*, vol. 8 (1977), pp. 77-101; J.E. McGuire, "Newton on Place, Time and God: An Unpublished Source," *The British Journal for History of Science*, vol. 11 (1978), pp. 114-129; Dobbs, *Janus Faces*; Schechner, "Newton and the Ongoing Teleological Role of Comets."

[72] Pierre Kerszberg, "The Cosmological Question in Newton's Science," *Osiris* 2nd series, vol. 2 (1986), pp. 69-106, 82. On these issues, see also: Newton, "De Gravitatione," pp. 106, 144; and Martin Tammy, "Newton, Creation, and Perception," *Isis*, vol. 70 (1979), pp. 48-58.

[73] Iliffe, "Isaac Newton and the Political Physiology of Self," p. 451.

[74] Isaac Newton, "First letter to Bentley" in "Four Letters of Sir Isaac Newton to Doctor Bentley," I.

The further active participation of God (in the form of gravitational forces and other active forces) is needed in order to create and sustain the present distribution of atoms in space. The observed order of the solar system could not obtain if it had the product of "blind metaphysical necessity," Newton writes in the *Principia*, since such necessity "is certainly the same always and everywhere, [and] could produce no variety of things."[75] Any observed order is a sign of an intended choice, more so regarding such sophisticated systems as the solar system. On this Newton writes:

> I see nothing extraordinary in the Inclination of the Earth's Axis for proving a Deity, unless you urge it as a Contrivance for Winter and Summer, and for making the Earth habitable towards the Poles; and that the diurnal Rotations of the Sun and Planets, as they could hardly arise from any Cause purely mechanical, so by being determined all the same way with the annual and menstrual Motions, they seem to make up the Harmony in the System, which as I explained above, was the Effect of a Choice rather than Chance.[76]

This passage makes it clear that Newton distinguished between the laws of motion and gravity[77] and the many possible orders that can be given to the material world depending on the chosen initial conditions. One question, which Newton did not resolve, was the mechanism through which gravitational forces operate. In the letters to Bentley, he argues that gravity cannot be an innate and essential property of matter, as is *vis insita*. His reasoning is the following:

> That Gravity should be innate, inherent and essential to Matter, so that one Body may act upon another at a Distance thro' a *Vacuum* without the Mediation of anything else, by and through which their Action and Force may be conveyed from one to another, is to me so great an Absurdity, that I believe no Man who has in philosophical Matters a competent Faculty of thinking, can ever fall into it.[78]

B. Cohen with R.E. Schofield (eds.), *Isaac* Newton's *Papers and Letters on Natural Philosophy and Related Documents*, (Cambridge, Mass., 1958), pp. 282-3.

[75] Newton, *Principia*, p. 442.

[76] "Newton to Bentley," Letter I, *Isaac* Newton's *Papers and Letters*, p. 236; *The Correspondence*, vol. 3, p. 236.

[77] What I referred to above as the difference between the infrastructure and software of the computer and its many computer programs.

[78] "Newton to Bentley," Letter III, *Isaac* Newton's *Papers and Letters*, p. 302-3; *The*

According to David Kubrin, Newton was not certain about the extent of God's role in the mechanism of gravity during the years 1692 to 1706,[79] since writing to Bentley:

> Gravity must be caused by an Agent acting constantly according to certain Laws; but whether this Agent be material or immaterial, I have left to the Consideration of my Readers.[80]

Nonetheless, Newton did say a few words about this Agent in the first letter to Bentley, where he remarks that "so great a Variety of Bodies, argues that Cause not be blind and fortuitous, but very skilled in Mechanicks and Geometry."[81] Thus from the correspondence with Bentley we can assume that Newton maintained the "Agent acting constantly according to certain Laws" is one "very skilled in Mechanicks and Geometry." This conclusion is also present in the background of the laws of motion and the definition of the absolutes of the *Principia*, which are mathematical, mechanical and constant. Yet, the laws of motion and the absolutes do not explain why we find a certain order in nature, since these laws are but the infrastructure upon which God may choose to impose any worldly order He wishes. In a letter to Thomas Burnet from 1680, Newton says this very clearly: "where natural causes are at hand God uses them as instruments in his work, but I doe not think them alone sufficient for ye creation."[82] If God uses the natural laws as instruments for creating a certain order it appears that a hierarchy of choices of differing orders played a prominent role in creation. Gravity is but the second level order imposed upon the first level order, which defines the passive structure of matter and its relation to the absolutes. In this sense gravity cannot be inherent to matter; it is a secondary cause that God chose to add to the three laws of motion. God employs gravity as a secondary lawfulness inserted into passive matter so that an "Agent skilled in mechanics and geometry" could add to this two level structure. The two level orders define the design of the material world but they do not yet determine any specific order. The quality and orderliness of the worldly design demonstrate the designer's skill in mechanics and geometry. Thus the harmony of the solar system is not only a proof that this order was not made by blind necessity; it is also a sign of the work of a Deity very skilled in mathematics and mechanics.

Inserting the lawfulness of gravity into the three laws of motion is a wise and brilliant move of a Deity who hides partially the reasoning behind his

---

Correspondence, vol. 3, pp. 253-254.

[79] Kubrin, "Newton and the Cyclical Cosmic," p. 338.

[80] "Newton to Bentley," Feb. 25, 1692/3, *Isaac Newton's Papers and Letters*, p. 303.

[81] "Newton to Bentley," Letter I, *Isaac Newton's Papers and Letters*, p. 287; *The Correspondence*, vol. 3, p. 235.

[82] Newton, *The Correspondence*, vol. 2, p. 334.

providence. Newton saw himself as belonging to a tradition of a select few who believe that God concealed his providence from the majority of humanity, yet dispersed and hid signs of the wisdom of his plan in such a way that only the chosen wise and moral can grasp it directly.[83] The wise of the ancient world, in return, followed God's way and hid the deep meaning of the laws of nature, including gravity, from the vulgar:

> But the Philosophers [Pythagorean] loved so to mitigate their mystical discourses that in the presence of the vulgar they foolishly propounded vulgar matters for the sake of redicule, and hid the truth beneath discourses of this kind. In this sense Pythagoras numbered his musical tones from the Earth, as though from here to the Moon were a tone... when meanwhile Pythagoras beneath parables of this sort was hiding his own system and the true harmony of the heavens.[84]

It seems to me that this providential concealment is the reasoning behind gravity. Imposing the secondary set of restrictions of gravity upon passive matter still leaves much freedom for any agent skilled in mechanics and mathematics to manipulate the first order of restrictions upon passive matter. The combination of inertia and gravity is a mathematical and mechanical solution for creating stable systems around central forces such as we see in the solar system. The more skilled the agent in mechanics and geometry the wiser and more magnificent will be his choice and use of central forces. In the following letter to Bentley Newton says that the two sets of restrictions upon matter (inertia and gravity) still leave much place for choice since they do not limit the Creator regarding any specific order; on the contrary, they open up a huge spectrum for the orders of many other worlds. The fact that we observe the regularity of the solar system shows that a third order of choice was imprinted upon the secondary (gravity) which was also a choice imprinted upon the first order (inertia) and that the creator of this design is extremely wise and powerful:

> Were all the Planets as swift as Mercury, or as slow as Saturn or his Satellites; or were their several Velocities otherwise much greater or less than they are, as they might have been had

---

[83] See: Snobelen, "God of gods," pp. 204-208; Rob Iliffe, "'Making a Shew': Apocalyptic Hermeneutics and the Sociology of Christian Idolatry in the Work of Isaac Newton and Henry More," Force and Popkin (eds.), *The Books of Nature and Scripture*, pp. 55-88, on pp. 79-81; Snobelen, "Isaac Newton, Heretic," pp. 389-91

[84] Newton, "Classical Scholia," revisions to Proposition VIII of the *Principia*, cited in Jamie James, *The Music of the Spheres: Music, Science and the Natural Order of the Universe* (New York, 1993), p. 165.

> they arose from any other Cause than their Gravities; or had the Distances from the Centers about which they move, been greater or less than they are with the same Velocities, or had the Quantity of Matter in the Sun, or in Saturn, Jupiter, and the Earth, and by consequence their gravitating Power been greater or less than it is; the primary Planets could not have revolved about the Sun, nor the secondary ones about Saturn, Jupiter and the Earth, in concentrick Circles as they do, but would have moved in Hyperbolas, or Parabolas, or in Ellipses very eccentrick.[85]

Thus the law of gravity is a divine free choice for implementing a certain mathematical lawful design, which still leaves open many possibilities for a worldly order.[86] During the years 1692 to 1706 Newton was searching for a mechanism that would support the notion that the designed world was a matter of a divine choice. This may be the reason why in those years he associated gravity with the divine will. Kubrin argues that even up to 1706 Newton held that "all motion in the world arose either through the effects of active principles or by the dictates of a will, the latter in Newton's sense could be either the will of an individual influencing the movements of his own body or the will of the Deity who has power over all nature."[87] But what precisely does it mean that the divine and the human wills can initiate a set of instructions upon matter in such a way that motion is created in the material design? I understand this as a suggestion that the difference between God's will and the human will is only a matter of the degree of influence upon the material design. The more skilled the agent becomes in mechanics and mathematics the more efficient are his deliberate instructions. Besides, the wisdom of the divine will is so great that it never dictates a loss of motion to the world system; this would only be the consequence of a non-skilled human command. "The power of God's will," Kerszberg says, "is such that the process of creation is nothing more than the coming into being of embodiments of the will itself. This process provides for a unity that is ultimately responsible for the intelligibility of the will itself."[88]

For Newton the difference between the divine and human will in relation to passive matter is fundamental. A will must be free otherwise choice has no meaning. God's will, which is the highest on the spiritual scale, cannot

---

[85] "Newton to Bentley," Letter I, *Isaac Newton's Papers and Letters*, pp. 285-6; *The Correspondence*, vol. 3, p. 235.

[86] The analogy I suggested of the many ordered programs written one upon the other on the infrastructure and software of the computer would correspond to a first, secondary, third, and even fourth set of orders chosen by an Agent skilled in mechanics and mathematics.

[87] Kubrin, "Newton and the Cyclical Cosmic," pp. 338-9. Kubrin supports his conclusion from Newton's manuscript in Cambridge Univ. Library Add. MS 3970, fol. 619r.

[88] Kerszberg, "The Cosmological Question," p. 88.

be restricted by any set of laws of nature. The only semi-restrictions upon God in creation come from his immutable and equable flowing absolute space and time, which are his internal qualities. Therefore the absolutes are not a matter of choice but are the qualities of the Creator, yet God is still free to choose not to utilize them in any sense. All other sets of laws are a matter of choice. For Newton, God chose to endow matter as well with the three qualities of his absolutes, that is, he gave material bodies his mathematical nature, the immutability of space, and the equable flow of time. Thus atoms are mathematical quantities that are immutable and indestructible, whilst the passive forces of atoms echo the equable flow of God's eternal time. Yet God's will is different from his absolutes, since nothing restricts it. If God had wished, matter could have had a different passive constitution. God chose the constitution he did because this choice was brilliant in its simplicity. Without further ado God sustains creation through His enduring presence.[89] Matter was designed specifically in accordance with God's eternal absolutes. Similarly, God's choice to add a second order of lawfulness to matter, namely gravity, is a proof of his wise, loving and just nature. Gravity is an instruction, or a commandment, so to speak, given to all bodies, obliging them to have an unconditional mutual mathematical relationship with all of their material "peers" in proportion to their distance and mass. This lawfulness is also a high form of justice since it is governed by the fairness of the third law of motion. Nothing forced God to choose such just lawfulness; this law is a sign of his wisdom, justice and love.

Newton's reasoning is not as explicit as this and does not openly confound physical laws with moral laws, yet in all his published works and his manuscripts such analogies are hinted at. For example, in the first text cited below, he says explicitly that there is nothing necessary in the laws of nature and in the following group of citations he associates the two original commandments with the laws of nature:

> And since Space is divisible *in infinitum*, and Matter is not necessarily in all places, it may be also allow'd that God is able to create Particles of Matter of several Sizes and Figures, and in several Proportions to Space, and perhaps of different Densities and Forces, and therefore to vary the Laws of Nature, and make Worlds of several sorts in several Parts of the Universe. At least, I see nothing of Contradiction in all this.[90]

> [Noah's religion] may be therefore called the Moral Law of all

[89] The metaphor of the electrically – operated computer again illustrates the role of God's absolutes in creation; they are the eternal current that is necessary for the running of the historical time of the worldly design.

[90] Newton, *Opticks,* pp. 403-4.

nations.[91]

> Thus you see there is but one law for all nations the law of righteousness & charity dictated to the Christians by Christ to the Jews by Moses & to all mankind by the light of reason & by this law all men are to be judged at the last day.[92]

> These [the moral laws of the true religion] are the laws of nature, the essential part of religion which ever was & ever will be binding to all nations, being of an immutable eternal nature because grounded upon immutable reason.[93]

It is not a coincidence that in these scattered notes Newton says that the moral realm exhibits one law of righteousness and charity for all nations and that this law is associated with the law of nature.[94] In both cases, the natural and the moral, laws restrict and bind all interacting objects. As gravity is a lawful restriction binding all material bodies within a universal mathematical relation in proportion to their distance and mass, so with the moral law governing human beings. The law of righteousness and charity binds all human beings together, in accordance with the closeness of the relationship. Human beings are more obligated first to their family, then to their friends, followed by the community, city, nation, etc., as in gravity the binding is proportionate to the distance. Whoever does not follow the moral law is punished according to a just calculation on the last day, as in a physical interaction the outcome of an impressed force on the interacting system is a proportional loss of motion.

---

[91] Newton, Keynes MS. 3, p. 27; cited also in Westfall, *Never at Rest*, p. 820.

[92] Newton, Keynes MS. 7, p. 3; cited also in Westfall, *Never at Rest*, p. 822.

[93] Newton, Yahuda MS. 15.5, f. 91; cited also in Westfall, *Never at Rest*, p. 821.

[94] This idea is expressed by the ancient Jewish philosopher Philo, who Newton carefully read whilst developing his idea of God's omnipresence. Philo writes: "While among other lawgivers some have nakedly and without embellishment drawn up a code of the things held to be right among their people, and other dressing up their ideas in much irrelevant and cumbersome matter, have befogged the masses and hidden the truth under their fictions, Moses, disdaining either course, the one as devoid of the philosopher's painstaking effort to explore his subject thoroughly, the other as full of falsehood and imposture, introduced his laws with an admirable and most impressive exordium.... [his exordium] consists of an account of the creation of the world, implying the world is in harmony with the Law, and the Law with the world, and that man who observes the law is constituted thereby a loyal citizen of the world, regulating his doing by the purpose and will of Nature, with accordance with which the entire world itself is also administered. "On the Account of the World's Creation Given by Moses," *Philo*, F.H. Colson and G.H. Whitaker (eds.), vol 1 (11 vol's; London, 1949), pp. 7-9. On Philo's influence on Newton, see: De Smet and Verelst, "Newton's Scholim Generale."

As mentioned in the previous section, in the decade following 1706, Newton changed his mind regarding God's role in the daily sustenance of gravity, and began to commit himself to an aethereal mechanism. In those years he was occupied with developing the notion of a natural mechanism by which "God could reform the system of sun and planets and renew the active principles in the cosmos"[95] with the aid of comets, to which Newton gave a central role in the replenishment of vapors and spirits lost by the emissions of the sun and the planets. Like many English theologians of the time, Newton believed that the cosmos, once left to itself, must decline in motion and regularity.[96] He remained undecided regarding which mechanism was responsible for this loss of motion of the solar system. Was it the will of providence? Or the will of human beings, who tend to follow their own corrupt dictates? Or was it a natural cyclical tendency of the material design to lose motion as time passed?

The Newtonian system, it seems to me, exhibits the same cyclical development on all three levels - the material, the human and the hidden work of providence. In the Bentley letters and in the second edition of the *Principia*, Newton insists that God periodically restores the quantity of motion and the regularity of movement among the celestial bodies by means of comets. The vapors from the tails of comets replenish the earth, the planets and even the sun.[97] The replenishment of the cyclical cosmos is another important indication of God's dominion and benevolence towards creation. The difference between these three levels –material, human and divine - is that the material design is totally dependent on the constant daily sustenance and the interactions taking place within the world, which include the actions of human beings. In the last query of the *Opticks*, Newton writes that by the passive principle of *vis inertiae* alone "there could have never been Motion in the World. Some other Principle was necessary for putting Bodies into Motion; and now that they are in Motion, some other Principle is necessary for conserving the Motion."[98] In other words, at no moment in the history of the cosmos can the material design be considered a closed system, as physicists believe today. Active forces, such as gravity, are essential for the running of the design and the seeming stability of the solar system. The tendency of matter towards disintegration requires a constant daily ordering, since:

> [I]f it were not for these [active] Principles, the Bodies of the Earth, Planets, Comets, Sun, and all things in them, would grow cold and freeze, and become inactive Masses; and all Putrefaction, Generation, Vegetation and Life would cease, and

[95] Kubrin, "Newton and the Cyclical Cosmic," p. 339.

[96] Markley, "Newton, Corruption and the Tradition of Universal History."

[97] On this, see: Kesrzberg, "The Cosmological Question," and Kubrin, "Newton and the Cyclical Cosmic," and Schechner, "Newton and the Ongoing Teleological Role of Comets."

[98] Newton, *Opticks*, p. 397.

the Planets and Comets would not remain in their orbs.[99]

Moreover, even the sophisticated mechanisms of gravity and other active forces are not enough for assuring the stability of the worldly design. Though God constantly sustains creation he needs to intervene periodically, since the bodily structures of the design, once created, do not remain constant but keep on undergoing cyclical processes differing in tempo. Newton's physical understanding of the cyclical cosmos is echoed in his studies of the works of the prophets and the history of ancient kingdoms in particular and humanity in general.[100] On all three levels a cycle of regeneration and degeneration is present due to the polarity between the passivity of matter and the active formative forces designing it. Providence is the active formative pole in charge of the regeneration and replenishment of the material cycle, and God's eternal flow is in charge of the constant sustenance of the material design of providence. Matter exhibits the passive pole, which has a natural tendency towards disintegration and degeneration.

The human dimension shares the tendencies of both the divine and the material. As long as human beings follow the commandments of the simple religion they help God in the sustenance of the design. The wise even help him in the regeneration of metals, as in alchemy.[101] Yet most people, according to Newton, are vulgar. They connect directly with sensible objects (the material pole), adulterating their senses with vice, and their worshiping assumes the form of idolatry. Idolatry is a materialistic substitute for the true spiritual purity of the simple creed. Once idolatry sets in human beings can no longer be attuned to God or to the mathematical divine music of the spheres,[102] since idolatry worship is a form of "lusting after material achievements" and "quarreling over metaphysical points."[103] Furthermore, idolatry distorts the direct connection of human beings with God, as the following manuscript suggests:

> And therefore Moses to prevent the spreading of this sort of Philosophy [the idolatry of the heathens, such as transmigration of souls] among the Israelites wrote the history of the creation

[99] Newton, *Opticks*, pp. 399-400.

[100] On this see also: Kenneth J. Knoespel, "Newton in the School of Time: The Chronology of Ancient Kingdoms Amended and the Crisis of Seventeenth-Century Historiography," *The Eighteenth Century*, vol. 30, no.3, (1989), pp. 19-41.

[101] The role of alchemy in the system is discussed in the following chapter.

[102] Penelope Gouk, "The harmonic roots of Newtonian Science," J.Fauvel, Raymond Flood, Michael Shortland and Robert Wilson (eds.), *Let Newton be! A New Perspective on his Life and Works*, (Oxford, 1988), pp. 101-125; Piyo Rattansi, "Newton and the wisdom of the ancients," pp. 185-201; McGuire and Rattansi, "Newton and the Pipes of Pan."

[103] Goldish, *Judaism*, p. 11.

> of the world in a very different manner from the Cosmogenies of the heathens, attributing the production of all things to the immediate will of the supreme God. Yet the Israelites by conversing with the heathens frequently lapsed into the worship of their Gods, & by consequence received their Theology, until they were captivated for these transgressions.[104]

Newton maintained that a real connection existed between idolatry and corrupt natural philosophy.[105] The more corruption enters the human dimension the more complex, vulgar, and unintelligible does natural science become. "It was not strange, therefore for Newton," P.M. Rattansi writes, "to try to prove that his scientific work in the *Principia* was a rediscovery of the mystical philosophy which had passed to the Egyptians and the Greeks from the Jews. It now made it possible to glimpse the truth hidden in ancient riddles, like the *music of the spheres* and the *Pipes of Pan* to whose music the whole of creation was said to dance.... Once human beings comprehended the infinite power of God, and how He had framed things and continually watched over them, they would be led to a deeper understanding and acceptance of the duties they owed to Him and to their fellow human beings."[106]

In both the theological and the natural domains, the divine pristine reform involved the recovery of the *prisca sapientia*—the original, primitive religion and the ancient knowledge of natural philosophy and mathematics. Thus for Newton, the misinterpretation of prophecy had "dire consequences for an individual's personal salvation."[107] Similarly, according to Markley, Newton's historical studies attest that the mathematical form of the prytanaea (the eternal flame) is a symbolic embodiment of the mathematical laws of the universe, which remains "a timeless guarantee of a divine intention, a design unaffected by the successive cycles of decay and the corruption of the material world." And precisely this sacred fire "offers the promise of access to the continuing bounty of Nature as a sign of God's grace."[108] Newton points out the ways that "religious ritual and religious worship enacts what is simultaneously the study and worship of the universe."[109] He writes:

> So then was one design of ye first institution of ye true religion to propose to mankind by ye frame of ye ancient Temples, the study of the frame of the world as the true Temple of ye great

---

[104] Newton, Yahuda MS. 15, p. 137.

[105] Manuel, *The Religion of Newton*, p. 42.

[106] Rattansi, "Newton and the wisdom of the ancients," pp. 198-199, his emphasis.

[107] Force, "Newton, the Lord God of Israel," pp. 131-58, 135.

[108] Markley, "Newton, Corruption and the Tradidtion of Universal History," pp. 136, 137.

[109] Knoespel, "Interpretive Strategies," p. 197.

> God they worshipped. And thence it was yt ye Priests ancinetly were above other men well skilled in ye knowledge of ye true frame of Nature and accounted it a great part of their Theology."[110]

These ancient priests were also natural philosophers who understood the relationship between the laws of nature and the moral laws. Newton's mathematical method of fluxions has a special status in such a system, proposing to present in a modern form the divine knowledge concealed in the ancient ceremonies of the priests around the sacrificial fire.[111] The gist of the method of fluxions is its mathematical technique, which will assist humans to connect directly with God's absolutes instead of remaining caught up in the material maze. When human beings interact directly with material objects confusion and vulgarity enter their human sensorium. In contrast, the method of fluxions calculates distortions undergone by equable flowing quantities, each calculation restoring momentarily the pure divine equable flow from which the flux of the distorted quantity has deviated; therefore, once it is embodied in physics this method enables human beings to cleanse their senses momentarily and converge finitely, with God's absolutes. I would like to suggest that for Newton, the use of this method in physics is analogous to an epistemological process of purification and detachment from bodily deviations. In a sense, the method of fluxions is a process of the restoration of the pure spiritual state of the sensorium before bodily distortions entered the sense organs, analogous to the restoration of sacrificial fire worship by chosen messengers after a corrupt period of idolatry.[112]

V. NEWTON'S WRITINGS ON PROPHECY

Newton was a millenarian and wrote extensively on the prophecies of Daniel and Revelation. The interpretation of the books of Daniel and Revelation formed one of the largest intellectual projects of the age, especially among Protestant scholars at Cambridge.[113] Newton was a heir to a continuous

---

[110] Newton, Yahuda MS. 41 f.7r., cited in Knoespel, "Interpretive Strategies," p. 196.

[111] See: "And in the Vestal ceremonies we can recognize the spirit of Egyptians who concealed the mysteries that were above the capacity of the common herd under the veil of religious rites and hieroglyphic symbols." Newton, Add MS 3990, f. 1, a passage from the early Book II of the *Principia* taken directly from "Origins of Gentile Theology," also cited in Westfall, *Never at Rest*, p. 434.

[112] Ramati, "The Hidden Truth of Creation," pp. 434-38.

[113] On millenarianism and the interpretation of prophecy of the Mede school, see: Steve Snobelen, "William Whiston: Natural Philosopher, Prophet, Primitive Christian," Ph.D. thesis, University of Cambridge, chapter 4, "Whiston and Prophetic Certainty"; Richard H. Popkin, (ed.), *Millenarianism*

tradition, dating from Joseph Mede in the early seventeenth century, followed by Henry More's published work on the two prophecies, and Ralph Cudworth's vast treatise on Daniel. This tradition saw "the growth of scientific knowledge as part of the fulfillment of the prophecies about events leading up to the actual reign of Jesus Christ on earth for one thousand years."[114] Joseph Mede, who can be considered as the founder of the tradition, was a professor of Greek at Cambridge and the author of *Clavis Apocalyptica* (1627). He believed that the beginning of the millennium was about to come, since Daniel had predicted that knowledge of the natural world would increase as the great event approached. Mede's students and disciples, such as Henry More, John Milton, Samuel Hartlib and John Dury started preparing for the new world to come by seeking to reform education and knowledge through planning new institutions and new theories. The "outburst of intellectual energy and innovation produced by Mede's reading of the import of the prophecies is amazing."[115] By the time Newton appeared on the scene, he was concerned both with the scientific problems raised by such men, and with the interpretation of the prophecies being studied by his teacher and friend, Henry More. Throughout his life, Newton worked on the scientific issues and the interpretation of the prophecies as discussed by More.

According to Frank Manuel, Newton believed that the "prophet was a religious teacher who had been favored and chosen by God because of his hard-won rational perfection, not his unbridled flights of fantasy." Though the language of prophecy was obscure and veiled, the mind of the prophet was pellucid in its clarity. "The meaning of prophecy was concealed," writes Manuel, "as were the laws of Nature, that other book in which God had written a record of his actions; and Newton drew frequent parallels between unraveling the mysteries of the books of prophecy and discovering the secrets of the Book of Nature."[116] More precisely, in Newton's view, prophecy, as well as the history of mankind, "becomes the analog of his natural philosophical efforts to uncover the means to reclaim humankind from corruption."[117] Thus Newton writes on Revelation:

---

and Messianism in English Literature and Thought 1650-1800 (Leiden, 1988); idem, "forward" in James E. Force, *William Whiston*; Katherine R. Firth, *The Apocalyptic Tradition in Reformation Britain 1530-1645* (Oxford, 1979); Paul Christianson, *Reformers and Babylon: English Apocalyptic Visions from the Reformation to the Eve of the Civil War* (Toronto, 1978); Bryan W. Ball, *A Great Expectation: Eschatological Thought in English Protestantism to 1660* (Leiden, 1975); Francis Bacon also called for a new rigorous science of prophecy. Francis Bacon, *The Works of Francis Bacon*, J. Spedding, R.L. Ellis, and D.D. Heath (eds.), (London, 1859), vol. 3, p. 341.

[114] Popkin, forward to Force, *Whiston*, p. xi.

[115] Popkin, forward to Force, *Whiston*, p. xii.

[116] Frank Manuel, *The Religion of Isaac Newton*, p. 88.

[117] Markeley, "Newton, Corruption and the Tradidtion of Universal History," p.135. For a further discussion on these issues see: Dobbs, *Janus Faces*; Mandelbrote, "A Duty of the Greatest Moment."

> Truth is ever to be found in simplicity, and not in the multiplicity and confusion of things. As the world, which to the naked eye exhibits the greatest variety of objects, appears very simple in its internal constitution when surveyed by a philosophic understanding, and so much the simpler by how much the better it is understood, so it is in these [prophetic] visions. It is the perfection of [all] God's works that they are all done with the greatest simplicity. He is the God of order and not of confusion. And therefore as they that would understand the frame of the world must endeavor to reduce their knowledge to all possible simplicity, so it must be in seeking to understand those visions. And they that shall do otherwise do not only make sure never to understand them, but *derogate* from the perfection of the prophecy.[118]

Here as elsewhere, we find the same central metaphor of truth as pure and simple, whereas deviation from truth is derogation from perfection. According to Castillejo, in all of Newton's diverse studies it was his "peculiar art to reduce unbelievable complexity to almost unbelievable simplicity."[119] Simplicity characterizes both the moral realm of prophecy and the epistemological realm of natural philosophy. Maurizio Mamiani has identified strong analogies between Newton's sixteen "rules" of prophetic interpretation and his later four rules of reasoning of the third edition of Book Three of the *Principia*.[120] "God guarantees that both Scripture and Nature can be understood by the human mind," says Snobelen on Newton, "what is more, both God's Word and God's Works were given and made in such a way that they are at the fundamental level simple and uncomplicated, and thus both must be approached with the same method."[121] A philosophical understanding in physics restores simplicity to the understanding, since "Nature does nothing in vain," says Newton, "and more is in vain when less will serve; for Nature is pleased with simplicity, and affects not the pomp of superfluous causes."[122] It is thus that bodies' curved trajectories are understood to be a momentary disruption from the equable flow of their *vis insita*. The same disturbed and

[118] Isaac Newton, 'Fragments from a Treatise on Revelation', in Manuel, *The Religion of Isaac Newton*, Appendix A, p. 120. My emphasis.

[119] Castillejo, *The Expanding Force*, p. 22. On the similarity of simplicity in the scientific method and prophetic interpretation, see: Force, "Newton's God of Dominion"; Snobelen, "God of gods," pp. 198-200; Markley, "Newton, Corruption, and the Tradition of Universal History."

[120] Maurizio Mamiani, "The Rhetoric of Certainty;" idem, "To Twist the Meaning: Newton's *Regulae philosophandi* Revisited," *Isaac Newton's Natural Philosophy*, Jed Z. Buchwald and I. Bernard Cohen (eds.), (Cambridge, 2001), pp. 3-14.

[121] Snobelen, "God of gods," p. 199-200.

[122] Newton, *Principia*, p. 320.

confused existence of bodies can also be detected in the history of commentary on prophecy. Newton believed that the words of Scripture, especially those of a prophecy, are the direct and simple words of the Creator. Thus like *vis insita* of atoms, prophecy is direct and simple, whereas commentary tends to divert and corrupt the original meaning of God:

> Hence is it not from any real uncertainty in the Scripture that Commentators have so distorted it; And this hath been the door through which all Heresies have crept in and turned out the ancient faith.[123]

> For to frame false interpretations is to prejudice men and divert them from the right understanding of this book. And this is a corruption equipollent to the adding or taking from it, since it equally deprives men of the use and benefit thereof.[124]

Interpretation of prophecy, writes Kochavi, is "conditional on a spiritual condition and, hence, the nature of the faith of the exegete."[125] Within such a frame of mind, it is only a short leap to perceive human corruption as a deviation or disruption and confusion of God's simple truth:

> I could wish they [the Catholic Church] would consider how contrary it is to God's purpose that the truth of his religion should be as obvious and perspicuous to all men as a mathematical demonstration. Tis enough that it is able to move the assent of those which he has chosen; and for the rest who are so incredulous, it is just that they should be permitted to die in their sins. Here then is the Wisdom of God, that he hath so framed the Scriptures as to discern between the good and the bad, that they should be demonstrations to the one and foolishness to the others.[126]

In this passage we can trace once again the theme of mathematical purity and its law-abiding force with its close connection to theology. The text even situates mathematics on a higher level than theological beliefs that involve complexity and confusion. The dominant drive or ardor underlying all of Newton's diverse investigations is also clear from this passage. Chosen and

---

[123] Isaac Newton, 'Fragments from a Treatise on Revelation,' p. 119.

[124] Newton, 'Fragments from a Treatise on Revelation,' p. 114. My emphasis.

[125] Kochavi, "One Prophet Interprets Another," p. 105.

[126] Newton, 'Fragments from a Treatise on Revelation,' p. 124.

wise souls have a more direct connection with the God of dominion through their hard-won rational perfection. They can distinguish between the good and the bad in Scripture by being able to understand his demonstrations. Is it a coincidence that Newton uses the word demonstration in the same short paragraph to refer both to mathematical demonstrations that prove truth beyond doubt and to the ability of the wise to recover God's demonstrations from the writings of Scripture? The recurrence of the notion of simplicity in the previous text referring to both realms is also notable. The true interpretation of prophecy and the ability to demonstrate the mathematical principles governing the motion of bodies require similar skills. Let us keep in mind that Newton said: "it is the perfection of [all] God's works that they are all done with the greatest simplicity." The reason behind this is that "He is the God of order and not of confusion. And therefore as they that would understand the frame of the world must endeavor to reduce their knowledge to all possible simplicity, so it must be in seeking to understand those [prophetic] visions."[127]

Did Newton believe he was chosen by God, as prophets were chosen due to their rational endeavor, to discover and present the epistemological means for recovering the direct connection with God? If this was so, why was he so careful not to say it openly? Why was he so cautious in pointing out the many analogies he had perceived between mathematics and theology? Why was he so secretive? For example, why did he not say explicitly that whoever shall follow the scientific method of the *Principia* and *Opticks* will restore the ancient religion? Why did he hide the theological message of the absolutes in restoring God's equably flowing time? He went as far as to write that his method distinguished between the pure and simple truth inherent in Nature and the impure and complex deviations of other natural philosophies. Moreover, underlying the following passage one can trace a hidden moral rule similar to the third law of motion of the reciprocity of action and reaction:

> [Those who are] inconsiderate, the proud, the self-conceited, the presumptuous, the sciolist, the skeptic, they whose judgments are ruled by their lusts, their interest, the fashions of the world, their esteem of men, the outward show of things or other prejudices [will never be able to] discern the wisdom of God in the contrivance of the creation.[128]

Why did he not openly point out that in both the natural and the moral realms a similar system assures that for every action there is a reciprocal reaction? He does say:

> [T]hese men "whose hearts are thus hardened in seeing should

[127] Newton, 'Fragments from a Treatise on Revelation,' p. 119.

[128] Newton, 'Fragments from a Treatise on Revelation,' p. 123.

> see and not perceive and in hearing should hear and not understand. For God has declared his intention in these prophesies to be as well that none of the wicked should understand as that as the wise should understand, Dan: 12."[129]

> But the world loves to be deceived, they will not understand.... There are but a few that seek to understand the religion they profess.... And chose to profess that religion which in their judgment appear the truest.[130]

Following Snobelen's remarkable reconstruction of the General Scholium, pointing out the hidden agenda of Newton's theological beliefs inserted between the lines of the text so that only the wise could understand their implications, and Snobelen's further insights regarding Newton's heresy, I surmise that Newton intentionally left to his future readers the work of interpretation as the prophets did for his own generation:[131]

> The giving ear to the prophets is a fundamental character of the true church. For God has so ordered the prophecies, that in the latter days "the wise may understand but the wicked shall do wickedly, and none of the wicked shall understand." (Dan. XII. 9, 10.)[132]

Daniel says explicitly that only at the end of human history will prophecy become clearer to most human beings. Thus it seems reasonable to assume that Newton believed his own scientific and theological work would have the same fate as other prophecies, remaining misinterpreted until the last days of the world. "Newton is convinced," writes James Force, "that God possesses dominion over the wise who will understand as well as over the wicked who, through vanity and idolatry, will not."[133] I would add to this that in Newton's systematic way of thinking it is more than possible that he believed the chosen wise souls to be equipped with a more refined and pure sensorium and sense organs by means of which they can comprehend God's words and actions more simply and directly than can a wicked and corrupt soul, which has accumulated idolatry and vanity in its "Organs of Sense,"

---

[129] Newton, 'Fragments from a Treatise on Revelation,' pp. 123-4.

[130] Newton, *A Treatise on the Apocalypse*, Yahuda MS. 1 f.5r., cited in Knoespel, "Interpretive Strategies," p. 199.

[131] Snobelen, "God of gods," and "Newton Heretic."

[132] Newton, *Observations*, Part I, # xvi, p. 304.

[133] Force, "Newton's God of Dominion," p. 83.

polluting the sensorium with corrupt and materialistic attitudes. Newton is not explicit on this issue as I am, yet in a letter to James Harrington from 30 May 1698, he suggests that there might be a connection between the mechanism governing sensation and the general laws of nature:

> I am inclined to believe some general laws of the Creator prevailed with respect to the agreeable or unpleasing affections of all our *senses*; at least the supposition does not derogate from the wisdom or power of God, and seems highly consonant to the macrocosm in general.[134]

In any case, simplicity is the key for understanding God's design:

> It is the perfection of all God's works that they are done with the greatest simplicity… And therefore as they that would understand the frame of the world must endeavour to reduce their knowledge to all possible simplicity, so it must be in seeking to understand these [prophetic] visions.[135]

Those who follow their own dictates instead of God's direct commandments have become degenerate and polluted and can no longer perceive and understand God's simple plan. We find a similar message in St. Paul words, which Newton most likely have read:

> Ever since the creation of the world his invisible nature, namely, his eternal power and deity, has been clearly perceived in the things that have been made. So they [who became idolatrous] are without excuse; for although they knew God they did not honor him as God or give thanks to him, but they became futile in their thinking and their senseless minds were darkened. Claiming to be wise, they became fools, and exchanged the glory of the immortal God for images resembling mortal man or birds or animals or reptiles.[136]

One needs Newton's "Rules for methodizing the Apocalypse"[137] as well as the grand mathematical apparatus and the laws of motion of his *Principia* to understand the hidden simplicity of providence. How can deceitful

---

[134] Newton, *The Correspondence*, vol. 4, pp. 274-275.

[135] Newton, Yahuda MS. 1.1 f 12r.

[136] Rom. 1:21-3.

[137] Newton, 'Fragments from a Treatise on Revelation,' p. 119.

souls perceive or understand simplicity if they have deviated from the equable flow of God's Absolutes? But the refinement and purity of the sensorium of the adept, to whom Newton belonged, enables them to follow God's pure and simple prophetic messages, as well as to discover the true mathematical passive laws of the motion of bodies, and the acts of Providence in world history. Wise and moral souls can also more properly recover the truth hidden in alchemy:

> Alchemy tradeth not with metals as ignorant vulgar think... this philosophy is not of that kind which tendeth to vanity & deceipt but rather to profit & to edification inducing first the knowledge of God... so that the scope is to glorify God in his wonderful works, to teach a man how to live well, & to be charitably affected helping our neighbours.[138]

Westfall has pointed out that the "philosophical tradition of alchemy had always regarded its knowledge as the secret possession of a select few who were set off from the vulgar herd both by their wisdom and by the purity of their hearts."[139] As Snobelen says, for Newton "ancient wisdom [had] become corrupt, and Newton was determined to restore its pristine purity."[140] The ancients, Newton wrote, "practised a two-fold philosophy, sacred, and vulgar: the Philosophers handed down the sacred to their disciples through types and riddles, while the Orators recorded the vulgar openly and in a popular style."[141] What is important to my discussion is that the mechanism for the original "pristine purity" as well as its potentiality to be restored are already built into the sensorium of humans. Originally the "Organs of Sense" and the sensorium are pure and simple, but as they get caught up in the material maze they become contaminated and vulgar. Relative and absolute notions, as we have seen, play a similar role in both physics and theology. In physics:

> the vulgar conceive those quantities [space and time] under no other notions but from the relation they bear to sensible objects. And thence arise certain prejudices, for the removing of which, it will be convenient to distinguish them into absolute and relative, true and apparent, mathematical and common.[142]

---

[138] Newton copied this text in the preface to "Manna" that Mr F. showed him in 1675: Keynes, MS 33, f. 5(v), cited in Westfall, *Never at Rest*, p. 298.

[139] Westfall, *Never at Rest*, p. 298.

[140] Snobelen, "God of gods," p. 203.

[141] Newton, Yahuda MS. 16.2, f. 1r (translated from Latin by Snobelen, "God of gods," p. 205-6).

[142] *Principia*, Scholium p. 13.

Similarly, in theology:

> It has been necessary to distinguish absolute and relative quantities carefully from each other because all phenomena may depend on absolute quantities, but ordinary people who do not know how to abstract their thoughts from the senses always speak of relative quantities, to such an extent that it would be absurd for either scholars or even Prophets to speak otherwise in relation to them. Thus both the Sacred Scripture and the writings of Theologians must always be understood as referring to relative quantities, and a person would be labouring under a crass prejudice if on this basis he stirred up arguments about absolute [*changed to* philosophical] notions of natural things.[143]

Thus, for Newton, absolute quantities are not for the masses.[144] Indeed, their vulgar nature has corrupted their sensorium and understanding to such an extent that they can no longer detect the simplicity of the absolute notions. The mechanism of "pristine purity" is also instilled in the *vis insita* of atoms. In the material world mathematical purity operates even more straightforwardly than in human souls. My reading of Newton's scientific work suggests that *vis insita* is the direct channel through which communication between God and matter is made through his absolutes. It is through this mechanism that God can act without being reacted upon and re-instill the pristine order of the whole solar system, although material interactions keep on causing deterioration and corruption in the order of the worldly design. Similarly, in the history of humanity, God re-instills his simple creed in men through prophetic souls who communicate directly with him through their pure pristine spiritual sensorium. It is not a coincidence that God reveals to his priests also the mathematical principles governing creation, concealing them in a religious ceremony that suits the capacity even of the vulgar to grasp simplicity and mathematics. Thus the prophets' concealed message to humanity is the embodiment of God's truth whether it is presented in scientific mathematical terminology or in the language of prophecy. Yet God's revealed truth retains its simplicity, order, and purity only for short periods of history, since even amongst pure souls attachment to material things as well as metaphysical garbage in the long run causes confusion, distortion, and departure from the seed of truth. Thus humanity, like the material world, forgets the simple creed and follows the dictates of corruption and idolatry.

My previous argument that Newton considered the method of fluxion to be a direct method of purification of the sensorium does not appear to hold. In spite of the resemblance of the process of purification in the moral and

---

[143] *Isaac Newton, The Principia*, p. 36. The text was cited also at the beginning of the chapter.

[144] Snobelen, "God of gods," p. 206.

epistemological realms, he seems to have separated the two. The natural philosophical epistemological route by itself can not bring true salvation to a suffering and sinful soul, though Newton does refer to the "purity of Mathematical and Philosophical Truth."[145] For the majority of humanity the removal of sin depends entirely on God's dominion and will. God, not human souls, decides to which souls he will reveal his truth, as only God has dominion over the replenishing of the solar system. Yet the choice is never unjust. Truth is revealed to those adept few who throughout their lives prefer the divine goodness and wisdom to their own egoistic tendencies. Without divine intervention the nature of the interaction of souls and matter, as well as of atoms amongst themselves, will most probably bring about corruption and deviation from the original religion. Therefore, salvation for all depends upon God's decision, and it is more than likely that Newton believed that even the simple straightforward divine scientific truth he was revealing to humanity would become materialistic, confused, and eventually deviate from God's path, since other human beings would interpret his writings in an idolatrous mode.[146] Yet notwithstanding God's sole dominion, the method of fluxions, as well as the other Newtonian works in natural philosophy, might purify the epistemological condition of human souls, by bringing them closer to the equable flow of the Lord and Master of creation if only properly understood. Indeed, as Knoespel concludes, it "is entirely appropriate to compare Newton's role at such an interpretive juncture to Jesus for just as the cogency of Jesus' interpretation established a new religion, so the cogency of Newton's work could ground religion on renewed physical truths."[147]

[145] *Isaac Newton: The Principia*, p. 414.

[146] Indeed, according to Snobelen, Newton hid most of his beliefs from society since he did not think the time had yet arrived when people would be able to comprehend. Snobelen, "Newton, Heretic," p. 419.

[147] Knoespel, "Interpretive Strategies," p. 201.

CHAPTER VII

GOD'S INFINITE PERSPECTIVE ACCORDING TO LEIBNIZ

> "It is true here too that those who seek the kingdom of God find the rest on their own way." (Leibniz, "Tentamen Anagogicum," 1696, *Philosophical Papers and Letters*, p. 477.)

I. PRELUDE

In Leibniz we encounter an entirely different meaning of free will, scientific objectivity and rationality, closely related to his calculus and God's choice of the best order of existence (the best spatial and temporal orders). Here the notion of scientific objectivity is more intricate and multi-leveled than in Newton. Leibniz does not give preference, as Newton does, to one unique absolute epistemological perspective represented by an immovable divine space and an equably flowing eternal duration. For him space and time are orders of existence of spiritual beings. From the coexistence of these beings he deduces the authenticity of an infinitely rich and layered notion of objectivity, internally encompassed within all subjective perspectives of the world. Leibniz's God perceives the infinite directly, so that for Him all possible orders of existence are wholes, which are prior to their parts.[1] This means that the omniscient God perceives the law of succession of states (time) of all possible worlds directly and distinctly, not partially and confusedly through the parts (events), as humans do. The calculus is a human (not divine) device to deduce the whole (the law of succession and the law of continuity of the world) from the parts (our limited finite successive perceptions). Or in Leibniz's terms:

> The essential ordering of individuals, that is, their relation to time and place, must be understood from the relation they bear to those things contained in time and place, both nearby and far, a relation which must necessarily be expressed by every individual, so that a reader could read the universe in it, if he were *infinitely sharp-sighted*.[2]

But what precisely does it mean to become infinitely sharp-sighted,

[1] See Leibniz, "On Mind, the Universe, and God," 1675, Daniel Garber and Robert C. Sleigh (eds.), *De Summa Rerum: Metaphysical Papers*, p. 5.

[2] *G.W. Leibniz: Philosophical Essays*, p. 183. My emphasis.

like God? How can a human being reach to God's infinite perspective from his limited and finite perception? Newton's solution in the form of finite inertial systems that flow along God's equable duration erasing all subjective and materialistic deviations is not legitimate in the Leibnizian system, since it ignores the complexity and infinitely layered nature of our perspective reality. For insight into the infinitely sharp-sighted perspective of God we need to realize that:

> [T]he actual world does not remain in this indifference of possibilities but arises from the actual divisions or pluralities whose results are the phenomena which are presented in practice and which differ from each other down to their smallest parts. Yet the actual phenomena of nature are arranged, and must be, in such a way that nothing ever happens which violates the law of continuity... On the contrary, things can be rendered intelligible only by these rules [such as those governing the calculus], for they alone are capable, along with the rules of harmony or perfection, which the true metaphysics provides, of leading us to the reasons and intentions of the Author of things. The result of this very great multitude of infinite compositions is that we are finally lost and are forced to stop in our application of metaphysical principles, and of mathematical ones as well, to physics. Yet these applications are never in error, and when a miscalculation appears after an exact chain of inference, it is because we cannot adequately examine the facts and because there is an imperfection in our assumption. *We are even more able to advance in this application to the degree that we are able to make use of our thoughts of infinity, as our newest methods have shown us.*[3]

The calculus was such a new method, which enabled human beings to perceive the world in a more infinitely far-sighted way once miscalculations were removed. Leibniz, in contrast to Newton, believed that God's rationality is truly open to human inspection when its nature is deciphered. Human beings who are able to understand God's rationality will come to realize that the rationality of the Supreme Being morally obliges him to choose among possibilities the minimal and most determined order of succession.[4]

---

[3] Leibniz, "Reply to the Thoughts on the System of Pre-established Harmony," *G.W. Leibniz: Philosophical Papers and Letters*, p. 583. My emphasis.

[4] See especially, Leibniz, "On the Radical Origination of Things," 1697, *Die Philosophischen Schriften von G.W. Leibniz*, vol. 7, pp. 302-8 and *G.W. Leibniz: Philosophical Papers and Letters*, p. 487; "Correspondence with Clarke," 1715-16, *G.W. Leibniz: Philosophical Papers and Letters*, pp. 677-680, 682-84.

II. PERCEPTION AND APPERCEPTION IN MONADS

In *New Essays* Leibniz repeatedly criticizes John Locke for saying that consciousness and reflection are two different things. Instead, he argues that both consciousness and reflection occur when we pay attention to our inner perceptions,[5] since it is wrong to assume that "the only perceptions in the soul are the ones of which it is aware." This point is significant, for Locke did not take into account the fact that:[6]

> We always have an infinity of minute perceptions without being aware of them. We are never without perceptions, but necessarily we are often without awareness [apperception], namely when none of our perceptions stand out.[7]

This paragraph indicates an important distinction. Leibniz believed that God created human beings from the outset as infinitely folded creatures, conscious only of a finite limited portion of their mental existence. To Locke, a human being is born as a *tabula rasa*, gradually developing personality out of whatever is experienced (always remaining within finitude), whereas, Leibniz maintained that rational monads, though conscious only of a partial perspective of their perceptions, enfold the infinite within their unconscious perceptual reality.

Leibniz held that all substances, except God, have a multitude of unconscious infinitesimal perceptions.[8] Substances are unities expressing the multitude of the world they live in with infinite perceptions. The nature of perception is intricate; matter or mechanical reasons alone can never explain "perception and what depends on it,"[9] yet "there is nothing in things except

[5] Not all of Leibniz's scholars agree that for Leibniz consciousness and reflection are the same thing. See e.g.: Mark Kulstad, *Leibniz on Apperception, Consciousness, and Reflection* (Munich, 1991); Nicholas Rescher, *The Philosophy of Leibniz* (Englewood Cliffs, 1967), p. 126f; Robert McRae, *Leibniz: Perception, Apperception and Thought* (Toronto, 1976), pp. 34-35.

[6] Leibniz, *New Essays on Human Understanding*, P. Remnant & J. Bennett (eds.), (Cambridge, 1989), p. 116.

[7] Leibniz, *New Essays*, p. 161f; *New Essays*, p. 53. Leibniz makes a similar criticism against the Cartesians, who like Locke, "disregard perceptions which are not themselves perceived, just as people commonly disregard imperceptible bodies." "Principles of Nature and Grace," & 4, *G.W. Leibniz: Philosophical Papers and Letters*, p. 637; see also: Monadology, & 14, *G.W. Leibniz: Philosophical Papers and Letters*, p. 644.

[8] Perception is a species in the genus of expression or representation. In perception a multiplicity is expressed in a unity. For a more detailed analysis of this point, see Robert B. Brandom, "Leibniz and Degrees of perception," *Journal of the History of Philosophy*, vol. 19 (1981), pp. 447-479; and Robert McRae, *Leibniz: Perception, Apperception and Thought*, pp. 20-26.

[9] Monadology, & 17, *G.W. Leibniz: Philosophical Papers and Letters*, p. 644.

simple substances [i.e. monads], and in them perception and appetite."[10] Monads are these simple substances, which we can never observe directly in the phenomenal world, though without them nothing would exist. Monads differ from one another in the way each one's perceptions express the same world. All monads are graded on a scale of perfection (or a scale of being) according to the distribution of proportion between their confused/unnoticed and distinct perceptions, using the doctrine of infinitesimal minute perceptions.[11] They are ordered in hierarchy according to their degree of perfection.

God is at the top of the scale. He is omniscient. As the most perfect monad, he knows the infinite perfectly and expresses his nature solely with distinct perceptions.[12] As such he is incorporeal and the embodiment of rationality. The lowest monads in the scale are simple substances expressing their nature with simple minute, unconscious perceptions.[13] Most often Leibniz labels such monads inanimate beings. If a monad does not have any kind of conscious perceptions it expresses its being solely by being a passive matter unaware of its perceptive reality. Yet when any kind of awareness arises within the perceptions of a monad it will be said that it is active and alive. Souls and spirits lie between God and simple substances on the scale of being. Souls are not themselves organisms, but the dominant monads of animals. Souls have sentiment, which is "something more than simple perceptions" because their "perceptions [are] more distinct and accompanied by memory,"[14] yet spirits are rational beings expressing thoughts and reasoning in addition to simple perceptions and sensations. The thoughts of spirits are developed perceptions, which have become distinct and are accompanied by consciousness or apperception. In his later years Leibniz employed the following mirror metaphor to express the distinction between souls and spirits:

---

[10] Leibniz, *G.W. Leibniz: Philosophical Papers and Letters*, p. 537. See also, Robert M. Adams. *Leibniz: Determinist, Theist, Idealist* (New York, 1994).

[11] "But since each distinct perception of the soul includes an infinity of confused perceptions which envelop the entire universe, the soul itself does not know the things which it perceives until it has perceptions which are distinct and heightened. And it has perfection in proportion to the distinctness of its perceptions." "Principles of Nature and Grace," & 13, *G.W. Leibniz: Philosophical Papers and Letters*, p. 640. Robert Brandom points out how all substances can express the same world through their perceptions, yet differ in their degrees of perfection. Robert B. Brandom, "Leibniz and degrees of perception," pp. 460-64. For an analysis similar to the following, see: McRae, *Leibniz: Perception, Apperception and Thought*, p. 27.

[12] For "only [God] has a distinct knowledge of everything, [because] he is the source of everything." "Principles of Nature and Grace," & 13, *G.W. Leibniz: Philosophical Papers and Letters*, p. 640

[13] Monadology, & 19, & 21, & 24, *G.W. Leibniz: Philosophical Papers and Letters*, pp. 644-645.

[14] Monadology & 19, p. 644; cf. Mon &25, &26, &28, p. 645. Leibniz is not always clear and consistent regarding what distinguishes animals from human beings. On this issue, see: Kulstad, *Leibniz on Apperception, Consciousness, and Reflection*.

> Souls in general are living mirrors or images of the universe of created beings, while spirits are also images of divinity itself...capable of knowing the system of the universe... each spirit being like a little divinity within its own sphere.[15]

Spirits, as images of divinity, have more developed perceptions within themselves than souls. Their distinct perceptions help them "rise to *reflective acts*, which enable [them] to think of what is called 'I' and to consider this or that to be in [themselves]."[16] Moreover, animals "do not know what they are nor what they do, and being constantly unable to reflect, they cannot discover necessary and universal truths,"[17] since they are but "living mirrors or images of the universe of created beings."[18] Unlike spirits, souls do not develop distinct perceptions, thus no distinction occurs within their inner reality between a knowing subject and the perceived object. Yet spirits are "like a little divinity within its own sphere"[19] since they, like God, have within themselves a distinction between a "creator" and his "creation," a "knower" and "what he knows," a "perceiving subject" and its "object." Leibniz says this as early as 1676 in his "Paris Notes":

> [T]hought has both subject and object... Yet it seems that self-consciousness is something *per se*, namely, a state; for in self-consciousness subject and object are the same.[20]

Apperceiving, which is a high stage of awareness, develops when monads transform their obscure, unnoticed perceptions into distinct ones in a process similar to the calculus's differentiation. As differentiation uncovers the hidden algorithm governing a geometrical sequence through a series of reflections on the differences between its terms so does apperception transform the previous hidden infinitesimal regularity governing the sequential order of monadic states into noticed inferential relations. These distinct inferential relations, in turn, become a finite expression of the sequential lawfulness of the

---

[15] Monadology, & 83, p. 651; cf. Monadology, & 29, & 82, pp. 645,651; "Principles of Nature and Grace," & 14, p. 640.

[16] Monadology, & 30, p. 646.

[17] "Discourse on Metaphysics," & 34, *G.W. Leibniz: Philosophical Papers and Letters*, p. 325.

[18] Monadology, & 83, p. 651, cited above.

[19] Monadology, & 83, p. 651.

[20] "Selections From the Paris Notes," April, 1676, *G.W. Leibniz: Philosophical Papers and Letters*, p. 160. The Paris notes are problematic because they contain themes, which Leibniz abandoned in his later years. For example, in his mature philosophy, Leibniz did not maintain that self-consciousness intrinsically reveals the whole nature of the spirit, though he kept the idea that subject and object express similar structures in self-consciousness.

infinitesimal and finite transitions of perceptions. This sequential order is the notion of time, thus the more inferential relations are uncovered the more aware does the rational monad become of the lawfulness governing its world. Sensation and memory are partial awareness of this lawfulness.[21]

Apperception (as a previous perception) is always a direct awareness of a past perception. Therefore, a continuous interplay occurs between a direct perception of an earlier one, which in turn is a perception of an earlier one, and so on. But why should a perception ever develop into another perception? This, according to Leibniz, is the work of appetition. Appetition, like the algorithm of mathematical sequences, governs the transitions of perceptions, and guarantees that the perceptions of all monads will continuously change.[22] Appetition in a specific monad governs its "law of the series." This law of order[23] is a law of progression, which guarantees that (within each monad engaging in a rational enterprise) the perception doing the reflection will always be more developed (at least in one inference) than the perception reflected upon.[24] Developed appetitions give rise to this kind of knowledge. It is here that awareness resides as a potential, and where self-consciousness emerges as an infinite process of a consciousness reflecting upon a "living mirror... with an internal action."[25] Each perception that has developed into a distinct perception becomes part of the "knower." Thus the more perceptions a monad transforms into distinct perceptions the wider is the conscious awareness of the soul of itself and of the laws governing creation.

Development of perceptions requires a continuous dialogue between a reflective consciousness and an internal active "living mirror"; it is a dialogue within one's perceptive awareness.[26] In spirits this "living mirror" illuminates the last stage of development of the reflective consciousness (i.e. the developed appetition or apperception), which perceives itself in the mirror as a unity expressing a multitude. Through this process of a mirroring dialogue the

---

[21] For example: "Being itself and truth are not understood completely through the senses.... Something is thus needed beyond the senses, by which to distinguish the true from the apparent." "On what is Independent of Sense and Matter," 1702, *Die Philosophischen Schriften von G.W. Leibniz*, vol. 4, p. 502, *G.W. Leibniz: Philosophical Papers and Letters*, p. 549.

[22] "The perceptions in the monad arise from each other according to the laws of the appetites." "Principles of Nature and Grace," & 3, p. 637, ibid, & 2, p. 636; Monadology, & 15, p. 644.

[23] *Die Philosophischen Schriften von G.W. Leibniz*, vol. 4, p. 518, *G.W. Leibniz: Philosophical Papers and Letters*, p. 493.

[24] *Die Philosophischen Schriften von G.W. Leibniz*, vol. 2, pp. 262-65, *G.W. Leibniz: Philosophical Papers and Letters*, pp. 533-35.

[25] "Principles of Nature and Grace," & 3, *G.W. Leibniz: Philosophical Papers and Letters*, p. 637.

[26] In New Essays, p. 211 Leibniz rephrases Philalethes' (the imaginary Locke) words as follows: ".. For the expression that occurs in the soul is like what there would be in a living mirror." A monad is also called a living mirror "endowed with an internal action." "Principles of Nature and Grace," & 3, p. 637). Leibniz uses the expression "living mirror" only in relation to spirits; I am extending its meaning by applying it also to self-consciousness.

rational monad gains further understanding of its unifying lawfulness. In other terms, the reflective aspect of consciousness engages in an operation very similar to mathematical differentiation. Both operations uncover a hidden regularity. The first transforms unnoticed perceptions into inferential relations with the aid of an active internal mirror, thus enlarging the conscious part of the spirit; the second reveals the governing algorithm hidden among the noticed members of a series. It seems to me that Leibniz's curves and their differing orders of infinitangular polygons (expressed by a higher order of differentials) are a mathematical expression of his peculiar experience of the spontaneity of self-reflexivity. To return to the spiritual experience recorded in the early Paris Notes, occurring around the time Leibniz developed his calculus:

> The following operation of the mind seems to me to be most wonderful: namely, when I think that I am thinking, and in the middle of my thinking I note that I am thinking about my thinking, and a little later I wonder at this tripling of reflection.... [W]e sense within ourselves this intention of the mind [the infinite process of self-reflexivity], by which we are brought back within ourselves and suppress what is external.... [T]he perception of a perception to infinity is perpetually in the mind, and in that there consists its existence *per se*, and the necessity of its continuation.[27]

As I have mentioned earlier, this mental experience of the reflexivity of perceptions to infinity reminds one of the process of dematerializing (or refining and transforming) one's daily perceptions (as Leibniz says, one loses the external world in the process). A rational being starts the process of reflection with confused perceptions of his/her internal surroundings. Each consecutive reflection enables the spirit to purify the original confused perception until the person eventually loses the external world (the confusedness) and reflects only upon the pure activity of ongoing self-reflexivity, that is upon the inferential relations governing the transitions (which function like the algorithm of the mathematical sequences).

In this system, a monad develops into an ego with thoughts when it comes to know the "necessary or eternal truths... which makes the connection of ideas indubitable and their conclusions infallible."[28] At this stage of development "souls are capable of performing acts of reflection and of considering what is called 'I'."[29] This happens when a monad perceives through its active mirror that its earlier perception was deduced from previous

[27] Leibniz, *De Summa Rerum: Metaphysical Papers*, pp. 74-5

[28] "Principles of Nature and Grace," & 5, *G.W. Leibniz: Philosophical Papers and Letters*, p. 638.

[29] "Principles of Nature and Grace," & 5, p. 638.

perceptions according to a specific progressive law, that is, when an ego perceives its own spontaneous activity of self-reflection. Thus consciousness is a developed perception that has grasped the infinite reflective nature of percipient beings:

> [S]o perception of perception goes on perpetually in the mind to infinity. In it consists the existence of the mind *per se* and the necessity of its continuation.[30]

The process of transforming perceptual confusedness into distinctness is internally experienced as distinct appetitions dematerializing their own self-image with the aid of an internal active mirror.

III: THE CALCULUS: THE BEST CHOICE

To reiterate: Leibniz gave a geometrical representation to the notions of the calculus, expressing its operations by an infinitangular polygon approaching a curve. His differentials are a continuous process of unfolding, which uncovers the simple hidden regularity that governs the whole sequence. This process of unraveling what is eternally folded and hidden within the curve is investigated with the assistance of higher order differentials, each ranging over fewer variations than the original terms of the sequence. Thus Leibniz's curve and his reflective higher order differentials are a visual expression of the continuous and progressive process of the unfolding and refinement of the original curve. In the Leibnizian calculus each successive higher order differential is a more refined expression of the law hidden within the original curve. Higher order differentials express more clearly the internal regularity of the instantaneous increments or decrements of the quantity of the original curve. A higher order sequence of variables is a clearer image of the original curve's regularity, since:

> *ddx* is the element of the element, or the *difference of the differences*, for the quantity *dx* itself is not always constant, but usually increases or decreases continually. And in the same way one may proceed to *dddx* or d^3x and so forth.[31]

[30] "Selections from the Paris Notes," April 1676, *G.W. Leibniz: Philosophical Papers and Letters*, p. 161. Once again, it is wise to be cautious with the Paris Notes, because they are very experimental.

[31] Leibniz, "Monitum de Characteribus Algebraicis," *G.W Leibniz: Mathematische Schriften*, vol. 7, pp. 222-23, translated in Bos, "Differentials," p. 19.

We thus achieve insight into Leibniz's calculus once we perceive it as a method of dematerialization and refinement. We may say metaphorically, that the curve, by the end of the process of differentiation, perceives itself and its regularity as a maximally determined developmental progression. The process of the differentiation of curves continuously transforms the confusedness of variation into a distinct and determined regularity, each time according to a smaller set of differences. But why are the infinitesimal passages (the infinitangular polygon), which move along straight lines more aspiring and refined than the original curvature differences of the curve?

The process of differentiation enables us to realize the superiority of straight lines over curves, such as a circle. Leibniz believed that the regularity of the variations of a circle does not function as a maximally determined quantity since the infinite and the finite sections of a circle differ. The tangents (the differentials) of a circle, which investigate the infinitely small variations of the curve, are straight lines, whereas the finite sections of the circle, though all are similar to themselves, are always an arc on a circle. Thus the circle is less determined than the straight line, since in the line:

> any part of a straight line is also straight, and a line is everywhere internally similar, nor can two parts be distinguished by means of their extremities.[32]

Through differentiation, one perceives clearly that the fundamental figure, which is enveloped in many curves, is the infinitangular polygon, that is, the most determined path from point A to its infinitesimal neighboring point B, is represented by a straight line:

> Such an interval of maximum determination, that is, the minimum and at once the most conformal figure made by the intervening terms is the simplest path from one to the other; in the case of points it is the straight line, which is the shorter between nearer points.[33]

"There is always a principle of determination in nature," writes Leibniz again and again,

> which must be sought by maxima and minima; namely, that a maximum effect should be achieved with a minimum outlay, so to speak.... Therefore, assuming that it is ordered that there

[32] "The Metaphysical Foundations of Mathematics," 1714, *G.W. Leibniz: Philosophical Papers and Letters*, p. 671.

[33] "The Metaphysical Foundations of Mathematics," p. 667.

> shall be a triangle with no other further determining principle, the result is an equilateral triangle is produced. And assuming that there is to be *motion from one point to another without anything more determining the route, that path will be chosen which is the easiest and shortest.*[34]

The Euclidean straight line, representing this shortest most determined path, is the only figure that is similar both in its infinitely small variations and in its finite appearance. It always unfolds into itself when reflecting upon its finite and infinitesimal existence, since "a line is everywhere internally similar to itself."[35] Moreover:

> From two points something new results, namely the locus of all the points which are uniquely determined by their situation in relation to the two given points, that is, the straight line which passes through the two points.... Two points determine the simplest path from one to the other, which we designate as a straight line [this relation is referred often as "that of maximal determination"].[36]

Leibniz admired the dynamic aspect of motion along a straight line. From his perspective, a path between two points along a straight line has the desirable feature of maximizing existence with minimum expenditure. The Leibnizian God chose this path among other dynamic principles (such as a path along a curvature), since it had the most efficient regularity. The Euclidean straight path is an economical passage, which clearly exemplifies Leibniz's employment of the law of continuity:

> *When the difference between two instances in a given series or that which is presupposed can be diminished until it becomes smaller than any given quantity whatever, the corresponding difference in what is sought or in their results must of necessity also be diminished or become less than any given quantity whatever.* Or to put it more commonly, *when two instances approach each other continuously, so that one at last passes over into the other, it is necessary for their consequences or results (or the unknown) to do so also.* This depends on a more

---

[34] Leibniz, "On the Radical Origination of Things," 1697, *Die Philosophischen Schriften von G.W. Leibniz*, vol. 7, pp. 302-8, translated in: *G.W. Leibniz: Philosophical Papers and Letters*, p. 487. My emphasis.

[35] "The Metaphysical Foundations of Mathematics," p. 671.

[36] "The Metaphysical Foundations of Mathematics," pp. 669, 671-2.

> general principle: that, *as the data are ordered, so the unknowns are ordered also.*[37]

> From this arises also the *law of continuity* which I was first to formulate, according to which the law of bodies at rest is, as it were, a special case of the law of bodies in motion; the law of equals, as it were, a special case of the law of unequals; the law of curves, as it were a special case of the law of straight lines. This is always true when a transition is possible from a genus to its limit in a special case, which is its apparent opposite.[38]

The law of continuity assures that the simplest path will be a regularity, which completely determines both the infinitesimal paths and the finite transitions in the world, in curves as well as in straight lines. Such a path enables God to govern two seemingly opposing phenomena with the same law of motion: rest (which is infinitesimal motion), and (finite) motion. "Geometry is full of examples of this kind," writes Leibniz, "but Nature,

> whose most wise Author uses the most perfect geometry, observes the same rule: otherwise it could not follow any orderly progress. Thus gradually decreasing motion finally disappears in rest... so that rest can be considered as infinitely small motion or as infinite slowness... So the rules for rest... can in a sense be considered as special cases of the rules for motion.[39]

The law of continuity guarantees that rest and motion will be but two extreme transitions, differing only in their tempos, lying along one progressive continuum of states.[40] Once again, the regularity of a circular motion (or a motion along any other curve) will be considered as a special case of the laws of straight lines once its infinitesimal paths are analyzed. If that were not the case God would need to create different laws to govern the infinitesimal transitions and the finite paths of the world. Yet this would require continuous

---

[37] "Letter of Mr. Leibniz on a General Principle Useful in Explaining the Laws of Nature through a Consideration of Divine Wisdom," *Die Philosophischen Schriften von G.W. Leibniz*, vol. 3, pp. 51-55, *G.W. Leibniz: Philosophical Papers and Letters*, p. 351. See also, *G.W. Leibniz: Philosophical Papers and Letters*, pp. 397-8. Italics in original.

[38] "The Metaphysical Foundations of Mathematics," p. 671. Italics in original.

[39] Leibniz, "Critical Thoughts on the General Part of the Principles of Descartes," *G.W. Leibniz: Philosophical Papers and Letters*, p. 398.

[40] Leibniz, "Specimen Dynamicum," 1695, *G.W Leibniz: Mathematische Schriften*, vol. 4, pp. 234-54, *G.W. Leibniz: Philosophical Papers and Letters*, pp. 435-450.

intervention from God on behalf of either the infinitesimal order or the finite one and that would undermine his perfection:

> For example, if God had established that a planet must always go on its own in a line curved like an ellipse [that is, in contrast to an Euclidean path], without adding anything explicable that caused or maintained this elliptical movement [that is, if God would not have added any more simple laws to the minimal laws he has chosen], I say that God would have established a *perpetual miracle* and that it could not be said that the planet proceeded thus in virtue of its own Nature or following natural laws [as in the case of a straight path where both the infinitesimal and the finite paths are governed by the same regularity], since it is not possible to explain this, nor to provide a reason for such a phenomenon.[41]

Thus, it is only with Euclidean paths that static and dynamic laws become an expression of a one continuous and determined regularity:

> [S]ince only force and the effort arising from it at any moment exist (for as we have explained above, motion never truly exists), and every effort [the infinitesimal potential] tends in a straight line, it follows that *all motion is in straight lines, or compounded of straight lines.* Hence it not only follows that *whatever moves in a curve strives* [once again the infinitesimal tendency] *always to proceed in a straight line tangent to it,* but there also arises here, the *true notion of firmness.*[42]

Yet if indeed God chose such minimal laws which no longer require him to intervene in the affairs of the world (since everything infinitesimal and finite is determined by the same continuous laws) how is it that we so rarely find bodies moving along straight lines in the phenomenal world?

Here Leibniz and Newton shared ideas. Both believed that God endowed physical entities with an internal force. This force brought bodies closer to God. It guaranteed that physical bodies would flow along straight lines when they had direct connection with the Creator. However, such a divine bond can hardly be found in reality. Material interference causes difficulties.

---

[41] Leibniz, *Die Philosophischen Schriften von G.W. Leibniz,* vol. 4, pp. 587-88, translation by Ezio Vailati, "Leibniz and Clarke on Miracles," *Journal of the History of Philosophy,* vol. 33 (1995), pp. 563-591, p. 573. My emphasis.

[42] Leibniz, "Specimen Dynamicum," *G.W. Leibniz: Philosophical Papers and Letters,* p. 449, italics in the original.

Material interactions keep disturbing bodies' direct connection with God while forcing them to move along curvilinear paths. Here the similarity ends, since for Newton the force in charge of the connection was a passive one, that is, *vis insita*, whereas for Leibniz it was an active force governing the transition of all monadic states.

For Leibniz, materiality was a principle of resistance, which deflected the body's internal dynamic principle (generating linear paths) from its course. Materiality brought about confused perceptions and curved paths by resisting (obstructing) the simple regularity of the internal, developmental, and dynamic principle of all transitions of the world.[43] If rational beings wanted to have a more direct connection with God they had to transform the confusion in their perception resulting from materiality back into God's original distinct active principles.[44] The calculus was an essential method for such dematerialization.

IV: LEIBNIZ'S LAWS OF MOTION

Leibniz's laws of motion are internal laws regulating the movement of material bodies, and stand in opposition to Newton's laws of Nature which function as limiting cases conditioning physical interactions.[45] Leibniz presented his mature theory of forces in *Specimen Dynamicum* (1695).[46] In this paper he described four different varieties of forces, which underlie all the levels of reality of his system. He labeled them primitive or derivative and further divided them into active and passive counterparts. Thus the four sets of forces are constituted of:

a. Primitive active forces.
b. Primitive passive forces.
c. Derivative active forces.
d. Derivative passive forces.

Leibniz developed these forces mainly in response to Descartes' notion of geometrical extension. His difficulty with the Cartesian extension was that it treated bodies as if they were inert and indifferent to motion or rest.

---

[43] Leibniz, "Specimen Dynamicum," pp. 437, 449. This point is elaborated in the next section.

[44] On this issue, see: John Dewey, "Leibniz's New Essays Concerning the Human Understanding," *John Dewey: The Early Works, 1882-88* (London, 1975), pp. 342-73.

[45] On Newton's laws of nature as limiting cases conditioning physical reality, see: Amos Funkenstein, *Theology and the Scientific Imagination*, pp. 89-96.

[46] See, "Specimen Dynamicum," pp. 435-450. For a description of the development of Leibniz's thought regarding forces, see: Daniel Garber, "Leibniz: Physics and Philosophy," Nicholas Jolley (ed.), *The Cambridge Companion to Leibniz*, (Cambridge: Cambridge University Press, 1995), pp. 270-353.

Thus Descartes' purely geometrical physical extension could not account for the order/regularity and stability which bodies exhibit. In particular, Descartes did not pay enough attention to the fact that bodies resist changes imposed upon them from the outside and participate actively in preserving their own corporeality.[47] Taking this up, Leibniz designed his four sets of forces to account for the stability and regularity of corporeal bodies. His theory of forces stood in opposition to Cartesian dualism. In the Leibnizian system, forces were situated in both the material and the metaphysical/monadic realms; the difference among them was not one of substance but of degree. Leibniz states this as follows:

> It must be maintained in general that all existent facts can be explained in two ways – through a kingdom of *power* or *efficient causes* and through a kingdom of *wisdom* or *final causes*; that God regulates bodies as machines in architectural manner according to laws of *magnitude* or of *mathematics* but does so for the benefit of souls and that he rules over souls, on the other hand, which are capable of wisdom, as over citizens and members of the same society with himself, in manner of a prince or indeed of a father, ruling to his own glory according to the *laws of goodness* or of *morality*. Thus these kingdoms permeate each other, yet their laws are never confused and never disturbed, so the maximum of the kingdom of power, and the best in the kingdom of wisdom, take place together.[48]

Thus the varieties of forces, both physical and metaphysical, express various levels of one reality. The primitive forces express the metaphysical reality of monads, called the kingdom of wisdom, whereas the derivative forces apply to phenomenal bodies, the kingdom of power.[49] The active counterparts refer to substantial forms and absolute motion; the passive, to the resistance and inertia of materiality (see figure 1).[50]

---

[47] "Discourse on Metaphysics," & 21, *Die Philosophischen Schriften von G.W. Leibniz*, vol. 4, pp. 446-47 and *G.W. Leibniz: Philosophical Essays*, pp. 53-54.

[48] Leibniz, "Specimen Dynamicum," p. 442, italics in original.

[49] Robert Adams correctly points out that Leibniz occasionally uses the distinction between primitive and derivative entities when he discusses the "law of the series" and the regularity according to which some term in the series moves toward the next one. See: *Die Philosophischen Schriften von G.W. Leibniz*, vol. 2, p. 262 and *G.W. Leibniz: Philosophical Papers and Letters*, p. 533. However, in most cases the distinction applies to the two separate realms - the metaphysical and the phenomenal. See: Adams, *Leibniz: Determinist, Theist, Idealist*.

[50] In what follows, I shall refer only to the active forces.

Monads:	Primitive active force	Primitive passive force
	Substantial form	Matter
	Clear perceptions	Confused perceptions
Bodies:	Derivative active force	Derivative passive force
	Living and dead forces	
	Laws of motion	Resistance and Inertia

Figure 1

Nonetheless, within his active forces, Leibniz outlines a formulation equivalent to Newton's first law:

> Also, since only force and the effort arising from it at any moment exist (for as we have explained above, motion never truly exists), and every effort tends in a straight line, it follows that *all motion is in straight lines, or compounded of straight lines*. Hence it ... follows that whatever moves in a curve always strives to proceed in a straight line tangent to it.[51]

Leibniz locates this law of motion within bodies, naming them primitive and derivative active forces. In most texts he attributes primitive forces to metaphysical monads while in the case of derivative forces he distinguishes between two sorts of forces - those applying to monads and those applying to the phenomenal realm of physical bodies.[52] His primitive and derivative active internal forces guiding bodies to move in straight lines are very different from Newton's *vis insita*. They are not a passive force persevering in an equable flowing state unless interrupted, but active spontaneous developments regulating physical and metaphysical states. Leibniz's forces, like his algorithms of mathematical sequences, enfold the body's whole development (past and future alike) and express the lawfulness

[51] "Specimen Dynamicum," part ii, 1695, *G.W. Leibniz: Philosophical Papers and Letters*, p. 449, italics in original. For comparison, Newton's Law I reads as follows: "Every body continues in its state of rest or uniform motion in a right line unless it is compelled to change that state by forces impressed upon it." *Principia*, Book I, Law I.

[52] See Adams. *Leibniz: Determinist, Theist, Idealist,*

of its world, analogous to the way that the appetition of monads governs the transition of perceptions.

For Leibniz the laws governing our world are the best ones possible. The omniscient God chose to create our world with the most determined laws that provide the maximal coexistence of substances. The active law of motion cited above is central among these laws. It guarantees that if "there is to be motion from one point to another without anything more determining the route, that path will be chosen which is the easiest or shortest."[53] All bodies in the phenomenal realm express this law internally with their derivative active forces. In the real metaphysical realm, in the kingdom of wisdom, monads express a correlative law with their primitive active forces. Primitive active forces, internal to all monads, govern the transition of perceptions:[54]

> There is moreover, a definite order in the transition of our perceptions when we pass from one to the other through intervening ones. This order, too, we call a path. But since it can vary in infinite ways we must necessarily conceive of one that is the most simple, in which the order of proceeding through indeterminate states follows the nature of the thing itself, that is the indeterminate stages are related in the simplest way to both extremes.[55]

From Leibniz's perspective it is not a coincidence that the law governing the path of all perceptions in monads corresponds to the law of motion of all bodies. Both laws, though operating in two separate kingdoms, "permeate each other" and express God's choice to govern the best possible world. In the following citation, Leibniz exemplifies how these two sets of laws, the natural and the moral, permeate each other, whilst adding explicitly that if a point moves along a curve and not according to its internal tendency in the most determined path, this is due to confused perceptions:

> But if this point could begin to be by itself, it would continue, not in the pre-established curve, but in the straight line tangent to it.... And while the point can have by itself only a tendency to move in a line tangent to this curve, because it has no memory or presentiment, so to speak, the entelechy expresses

[53] *Die Philosophischen Schriften von G.W. Leibniz*, vol. 7, p. 304 and *G.W. Leibniz: Philosophical Papers and Letters*, p. 487.

[54] Leibniz does not explicitly associate the primitive active force and the transition of perceptions in this text. However, I maintain that he did connect the two.

[55] "The Metaphysical Foundations of Mathematics," *G.W. Leibniz: Philosophical Papers and Letters*, p. 671.

> the pre-established curve itself, since the surrounding bodies cannot have any influence upon this soul or entelechy, so that in this sense there can be no violence with respect to it. What is commonly called violence, however does occur insofar as this soul has confused, and therefore involuntary, perceptions.[56]

I understand this rather obscure text as saying that any resistance to following the shortest path from point A to point B is resistance to the laws of the kingdom of power instilled internally in the point. This resistance is also expressed in the kingdom of wisdom as confused perceptions of monads. The reasoning behind this regularity and confusion reflects the chosen divine laws regulating and restricting the path of monadic states, internal to monads and bodies. These laws guarantee that a maximum number of essences coexist, yet when anything resists this regularity the resistance will be automatically expressed in the phenomenal realm as materiality and in the metaphysical realm as confused perceptions. Both realms are instances of God's Principle of Determination, which always "prefers the maximum effect at the minimum cost."[57] The calculus dematerializes epistemologically the material resistance and transforms the confusedness into distinctness.

A further expression of Leibniz's distinctive minimal maximizing principles is his formulation of a law equivalent to Newton's third law of motion:

> [E]very passion of a body is spontaneous or arises from an internal force, through an external occasion.... From what we have said we can understand that the action of bodies is never without reaction, and both are equal to one another, and directed in opposite directions.[58]

Like Newton, Leibniz assumes here that action and reaction occur in pairs. He perceives this combination in the phenomenal world, but believes it has no influence on the relation between cause and effect. Causes and effects are "intelligible notions, which belong to the understanding... [and as such] are

---

[56] Leibniz, "Reply to Mr. Bayle's Critical Dictionary, Article Rorarius," 1702, *G.W. Leibniz: Philosophical Papers and Letters*, p. 577.

[57] *Die Philosophischen Schriften von G.W. Leibniz*, vol. 7, p. 304, translated in: *G.W. Leibniz: Philosophical Essays* p. 150.

[58] "Specimen Dynamicum," part ii, *G.W. Leibniz: Philosophical Papers and Letters*, pp. 448-9. For comparison, Newton's Law III goes as follows: "To every action there is always opposed an equal reaction; or, the mutual actions of two bodies upon each other are always equal and directed to contrary parts." *Principia*, Book I, Law III.

above the imagination [i.e. are not sensed]."[59] They are applied to phenomenal reality, the kingdom of power, through reasoning. Causes are distinguished from effects and "a preceding instant always has the advantage of priority, not merely in time but in nature, over following instants."[60] Thus, for Leibniz, in contrast to Newton's physics (not his metaphysical and theological studies), causes are the progenitors of effects, and not vice versa.

Leibniz adds conservation principles to causal events because "Nature never substitutes things unequal in their forces for one another... [so] the entire effect is always equal to the full cause."[61] All this reflects his belief that God predestined the entire inventory of a monad. Whereas Newton maintained that motion is always on the decrease in the world, therefore God intervenes and periodically replenishes creation, Leibniz held that motion is always conserved and God never intervenes in the doing of the world. Monads are an immanent closed system of a developing unity expressing a plurality. A specific law regulates everything that happens to the monad. No perception gets lost; everything is preserved. In the kingdom of wisdom, monads reflect upon perceptions and appetitions, which they continuously unfold and transform. More highly developed perceptions indicate that the self has gained a better understanding of the universal laws governing all its perceptions without, losing the content of its past perceptions. Dematerialization of perceptions means that the self discovers and conserves its own operations and essences from its unnoticed infinitesimal perceptions. All changes whether carried out in the kingdom of power or in that of wisdom are but transformation of states, everything is conserved.

It seems to me that the correlation between the perceptual transition of monads and the laws of motion of all bodies is not a coincidence. These corresponding general laws, distinctly expressed in the metaphysical and phenomenal realms, indicate that Leibniz was thoroughly convinced that the calculus revealed God's rationality and the governing laws of existence.

V. THE PURIFYING ROLE OF THE METHOD OF FLUXIONS AND THE CALCULUS

The great difference in the role of Newton's method of fluxions and Leibniz's calculus in the affairs of the world thus becomes more apparent. Leibniz's calculus was a model for a process of refinement of the perceptual reality of monads uncovering the true and determined laws of nature governing

---

[59] *Die Philosophischen Schriften von G.W. Leibniz,* vol. 6, p. 502, translated in: *G.W. Leibniz: Philosophical Papers and Letters,* p. 549.

[60] *Die Philosophischen Schriften von G.W. Leibniz,* vol. 3, p. 582, translated in: *G.W. Leibniz: Philosophical Papers and Lettres,* p. 664.

[61] "Specimen Dynamicum," part i, *G.W. Leibniz: Philosophical Papers and Letters,* p. 444.

all transitions in the world. With its aid he was able to dematerialize and purify perceptions and transform the confusedness resulting from materiality into the original distinct unity of God's rationality. The perceptual reality of the monad contained concealed information regarding the infinitesimal linear paths guiding all future finite paths of its monadic states. These most determined infinitesimal paths express distinctly the rationality of the continuous laws governing the world, which fully determine both the infinitesimal and the finite transitions of all monadic states. Leibniz believed that God perceives distinctly all these infinitesimal passages in the created world. God's infinitesimal and infinitely far-sighted perception enabled him to chose the simplest minimal continuous laws, which govern the maximal number of coexisting entities. God is the only monad that has distinct perceptions of all the transitions of the state of the world; all other monads' perceptions have a differing proportion of degrees of distinctness and confusedness.[62] Their awareness of the interconnectedness of everything is limited in so far as they still have confused perceptions.

God is wholly spiritual without a body (He is incorporeal), since the distinctness of his perceptions, that is, his pure spirituality, does not resist the rationality of the world. Thus "God alone is entirely detached from body."[63] Other monads resist God's rationality with a body, which is a finite, confused and folded expression of the lawfulness of the best order of existence.[64] God's rationality is the ultimate objective perspective of the world, yet each monad partially expresses this objectivity according to its capacity of proportion between confused and distinct perceptions. God's perspective is not a unique or concrete point of view, since God has no body. God's distinct perceptions become concrete and real only when he creates a world according to the best lawfulness. His rational perspective becomes objective with the realization of a created world populated with an infinite number of entities each expressing a unique perspective of that same lawfulness of the best order, according to each one's capacity. Therefore, it becomes imperative that each monadic perspective will be an essential parcel of the total objective perspective of the world. The true objective perspective of the world is always intrinsic to individual monads. It is expressed in multifarious degrees of clarity in monads since each monad has different degrees of confusedness and distinctness of God's rationality:

> Each soul knows the infinite, knows everything, but confusedly... our confused perceptions are the result of the impressions made on us by the whole universe. It is the same with each monad. Only God has distinct knowledge of

[62] "The Monadology," *G.W. Leibniz: Philosophical Papers and Letters*, p. 650.

[63] "The Monadology," p. 650.

[64] "The Monadology," p. 650.

everything, for He is the source of everything.[65]

The more clearly the lawfulness of the world is expressed in the perceptions of a specific monad, the more it expresses God's distinct rationality and the better is its ability to perceive its specificity and limited perspective in relation to others.

The essential contribution of each monadic point of view to the realization of the world may be exemplified with the aid of the calculus. Leibniz's calculus is a human tool intended to assist rational monads to dematerialize certain of their confused perceptions into distinct ones. The process of differentiation enables monads to purify their perceptions and gain insight into the hidden rational lawfulness of their world. The calculus reveals to rational beings that though God thinks the whole (the successive order, i.e. the law of continuity) prior to its parts, whereas they comprehend the lawfulness of their world only from the facts and episodes revealed to them throughout life, they all share the rationality of God. Monads might encompass a confused and partial perspective of the order of succession of their world, by perceiving only partial events and happenings. They might assume accordingly that the succession they perceive has no purpose and no relatedness to other beings, or worse, that there is no intrinsic relation between their past, present and future. But all this really means is that they are still confused. Becoming more rational is a continual process of transforming confused perceptions into distinct ones in a process similar to differentiation. A rational monad can grasp the hidden lawfulness governing the conscious episodes of its life by differentiating the unconscious undeveloped perceptions hidden between these finite episodes. Such growing awareness of the interconnectedness of past and present events will necessarily be accompanied by a further recognition of the remarkable ways in which God's minimal laws govern the harmonious coexistence of monadic states.

In contrast to Leibniz's intricate and multi-layered notion of objectivity, I understand Newton's scientific objectivity as converging finitely with God's absolute perspective, assuming that human beings can reach this perspective only from an external, finite, and limited position. Mortals will never be able to understand God's reasoning and choices or how he acts without being reacted upon, unless and until God chooses to reveal the whole truth to them. For this reason, Newton did not speculate on the nature of God's rationality but adopted the empirical method of analysis and synthesis in science and believed the word of prophecy. To come to know God's work in nature an analysis requiring empirical experiments and observations is crucial. On "the basis of this empirical starting point," writes James Force, "Newton inductively derives probationary 'Principles' such as the Inverse Square Law by "Analysis." The second part of his method is the synthetic deduction of

[65]"Principles of Nature and Grace," *G.W. Leibniz: Philosophical Papers and Letters*, p. 640.

future phenomena on the basis of these 'Principles.'"[66] This empirical method enabled Newton to explore God's providence in world history, as well as in nature. "In Newton's mind," writes Dobbs, "history seemed to bear a direct correspondence with experimental or even mathematical demonstration. For Newton it was only the firm correspondence of fact with correctly interpreted prophecy that provided an adequate demonstration of God's providential activity."[67] Newton believed that human beings can decipher God's work in history and in nature only through careful attention to the details. Only the wise will be able to decipher the laws governing these details and thus comprehend God's providence. The closer one comes to achieving a glimpse of God's truth, the simpler will be the method he will use since God's truth is always simple.

This difference is crucial. For Leibniz, a monad's perception which has become truly distinct is not an external convergence with God's objective perspective, but is itself a truly godly perspective. Humans and God share the same distinct rationality. What limits mortals is not a Newtonian incompetence to understand God's rationality and choices, but an inability to perceive all of the monad's infinite perceptions distinctly. Rational monads are not able to perceive distinctly all the infinitesimal passages of their world since confusedness still exists in their perceptions. Nonetheless, with the aid of the machinery of differentiation (as well as other Leibnizian methods) they can uncover some of the regularities of the successions they perceive in a finite and confused way. Unlike the Newtonian God who separates his existence and duration from his will, Leibniz's God is totally rational. The choice of the best order of existence is rational; it is the best one. For the Leibnizian God the true, the good and the beautiful are one and the same choice. Therefore, when rational monads have a partial objective perspective of their existence, it is not a finite external convergence with the divine absolutes in the Newtonian sense, but a finite internal perspective of the lawfulness of the best order of existence indistinguishable from God's rationality. The order of existence (the spatial and temporal order) of Leibniz's world is completely rational and intrinsic to all monads, whereas Newton's absolutes (time and space) are not rational but are the external condition, the limiting cases of the material world. The absolutes are the limiting case because God chose to endorse material atoms with the three attributes of his absolutes, but if he had wished he could have done otherwise. Nothing limits and restricts the will of the Newtonian God.

Furthermore, and this is again an important difference, Newton believed in a cyclical cosmology, whereas for Leibniz the world becomes more and more perfect as time goes on. Newton was convinced that the highest form of purity existed just before the Fall of Man. Noah's pristine religion, as well as all other religious restorations, were reminiscences of this simple state of truth. The material order of the design, as well as humanity itself, has gone through different stages of purity. Humanity began with a pure stage of worship, and

---

[66] Force, "Newton, the Lord God of Israel and Knowledge of Nature," p. 147. See also, *Opticks*, pp. 404-5.

[67] Dobbs, *Janus Faces*, p. 85.

gradually descended into a more and more corrupted moral stage through idolatry. Processes in nature as well as in humanity are always cyclical. Transformation and purification occur only when God personally intervenes in the affairs of the world. Real and complete transformation back to the purity of creation is not possible unless God chooses it. Leibniz, however, believed that the transformation and purification of monadic states is a progressive process of enlightenment internally folded into monads. "The entire universe is involved in a perpetual and most free progress, so that it is advancing toward greater culture."[68] No apocalyptic end or cyclical cosmos enter the law of the series of monads. The world progresses as monads' perceptions unfold.

What can we make of this difference between Newton's and Leibniz's understanding of God's ultimate perspective? Newton's absolutes appear to function as the objective divine reference of the created world and are closely associated with his method of fluxions. Thus his natural philosophy encourages people to aspire to an inertial state of consciousness, which can be interpreted as an epistemological process of purification of the senses obliterating deviations caused by material interaction governed by the laws of action and reaction. Newton's ideal of scientific objectivity denies personal differences by removing individual positions from the mesh of a troublesome material existence, and making them converge momentarily with an absolute mathematical divine inertial stage which does not participate in any material interaction, yet conditions the laws of motion of all bodies. This epistemological stage converges finitely with the eternal divine equable flow, becoming the embodiment of spiritual purity representing the highest form of detachment from material conditions. Someone who can momentarily acquire this detached and peaceful epistemological stage is able to go beyond the stage of action and reaction. For Leibniz, in contrast, objectivity is a multifarious, rich and layered perspective encompassing the internal perspective of each individual monad. The more a rational monad apprehends the conditions of its existence the better is it aware of its specific point of view in relation to the shared rationality of the chosen world. The more one becomes objective the more can one identify the rationality and regularities of the world without projecting his/her particularity upon others, while acknowledging the necessity of a pluralism of perspectives. For Leibniz, only God can be fully objective thus only He has no body.

---

[68] Leibniz, "On the Radical Origination of Things," *G.W. Leibniz: Philosophical Papers and Letters*, p. 491.

CHAPTER VIII

THE SACRIFICIAL FIRE AND ALCHEMY

"[In the Prytanaea ceremony] ye sacred fire was a type of ye Sun & all the elements are part of that universe wch is ye temple of God." (Newton, Yahuda MS. 41, f.8r.)

"When [the alchemical agent] is introduced into a mass of substances its first action is to putrefy and confound into chaos; then it proceeds to generation." (Newton, Keynes MS. 12, ff.1v-2.)

So far I have argued we have good reason to maintain that Newton's method of fluxions functions also as an epistemological method purifying the mind from material distractions. We have also gained insight into Newton's spiritual motives through a comparison with Leibniz's calculus. To pursue the meaning Newton assigned to the true worship around sacrificial fires we shall now look at further connections amongst his method of fluxions, Noah's original pristine religion, alchemy and the Jewish Temple.

As we have seen in previous chapters, Newton's studies of the history of the original religion were motivated by a desire to discover the ancient pristine natural philosophy. He believed that this knowledge could bring humanity closer to salvation. Indeed, as Markley writes, Newton's "universal history becomes the analog of his scientific efforts to uncover the means to reclaim humankind from corruption."[1] Purity lies at the heart of the Noachian religion. The eternal flame at the center of the altar represents the truth regarding the heliocentric solar system, or in Newtonian terms:

> So then was one design of ye first institution of ye true religion to propose to mankind by ye frame of ye ancient Temples, the study of the frame of the world as the true Temple of ye great God they worshipped. And thence it was yt ye Priests anciently were above other men well skilled in ye knowledge of ye true frame of Nature & accounted it a great part of their theology.[2]

---

[1] Markley, "Newton, Corruption, and the Tradition of Universal History," p. 135.

[2] Newton, Yahuda MS. 41, f.7r. cited also in Markley, "Newton, Corruption, and the Tradition of Universal History," p. 136.

It seems to me that Newton considered this knowledge of the "true frame of Nature" to consist of divine secrets essential to the sustenance of purity in a material world that deteriorates as time passes. What are those divine natural philosophical secrets? Markley assumes that the fire of the "prytanea functions as a kind of computational code by which true knowledge can be demonstrated and transmitted without interruption, interference, or corruption." The prytanaea, representing the heliocentric system, has the status of a *universal form*; nonetheless it "transcends its material incarnations." Though "ancient temples may be destroyed ... the knowledge they encode endures." The form of this divine wisdom "is a kind of mathematical proof that God is universal, abstract, indivisible, and all-powerful." Furthermore, and even more profoundly, the sacred flame "becomes a complex symbol of both moral renovation and technoscientific knowledge and, in both respects, can be identified with the purifying flame of the alchemical sublimation."[3] In comparison with the true worship around sacrificial fires, idolatry starts when worshippers become materialistic and confuse "form and content, the ideal and the material, the timeless and the corrupt." Indeed, "worshipping the mere representations of divine order – the Sun, stars, and planets – turns men and women away from technoscientific knowledge and true faith and makes them subject to self willed delusions."[4] Thus the corruption of the pristine religion, for Newton, occurs when:

> The frame of the heavens consisting of Sun Moon & Stars being represented in the Prytanea as ye real temple of the Deity men were led by degrees to pay a veneration to these sensible objects & began at length to worship them as the visible seals of divinity. And because ye sacred fire was a type of ye Sun & all the elements are part of that universe wch is ye temple of God they soon began to have these also in veneration. For tis agreed that Idolatry began in ye worship of ye heavenly bodies & elements.[5]

But how can one get access to the true knowledge of the pristine natural philosophy, which lies at the heart of the ceremony of the sacred fire, without falling into any form of idolatry? Markley argues that for Newton the sacrificial fire necessarily had a complex of meanings: "it is a spiritual light; a symbol of the divine wisdom manifest in the heliocentric solar system [and] the transforming flame of alchemy."[6] Accordingly, the key to the spiritual knowledge surrounding the sacrificial fire lies in alchemy and in encoding the

---

[3] Markley, "Newton, Corruption, and the Tradition of Universal History," pp. 136-7.

[4] Markley, "Newton, Corruption, and the Tradition of Universal History," p. 138

[5] Yahuda MS. 41, f.8r, Markley, "Newton, Corruption, and the Tradition of Universal History," p. 138, cited also in this book previously in chapter II.

[6] Markley, "Newton, Corruption, and the Tradition of Universal History," p. 137.

ancient natural philosophy of the harmony of the heavenly spheres. Dobbs too believes that Newton's true religion is "the worship of God for his activity in the world – in creating it, in preserving it, and governing it according to his will."[7] Therefore, a deep connection exists between the moral stage of humanity and the stage of natural philosophy. Newton writes at the end of the *Opticks*:

> If Natural Philosophy in all its parts, by pursuing this method, shall at length be perfected, the bounds of Moral Philosophy will also be enlarged. For so far as we can know by Natural Philosophy what is the first cause, what power He has over us, and what benefits we receive from Him, so far our duty towards Him, as well as that towards one another, will appear to us by the Light of Nature. And no doubt, if the worship of false gods had not blinded the heathen, their Moral Philosophy would have gone farther than to the four cardinal virtues; and instead of teaching the transmigration of souls, and to worship the sun and dead heroes, they would have taught us to worship our true Author and Benefactor, as their ancestors did under the government of Noah and his sons before they corrupted themselves.[8]

Thus the stage, methods and content of natural philosophy influence the moral stage of humanity, and vice versa. No true science can develop in a society that worships idols in any kind of form. The question remains open regarding what Newton understood to be the true pristine science. Dobbs reminds us that caution is needed in retracing the original pristine natural philosophy only with reference to alchemy. Entrance to the secret knowledge of alchemy opens the way to the truth of the original religion, yet one needs to bear in mind that, according to Newton, alchemy was the earliest form of corruption and ultimately obscured parts of the primitive true religion. Though indeed somewhat corrupted, alchemy still has much to say about the true pristine science.

The first thing that becomes clear from Newton's alchemical studies as well as his physics is that "matter [is] passive and that only the spiritual realm [can] initiate activity."[9] Though matter has fallen from its pristine state yet it is still possible to redeem it with the philosopher's stone. Alchemy aspires to unlock the secrets of matter and transform that which has become a degraded form of iron into an elevated form of gold. It is the natural

---

[7] Dobbs, *Janus Faces*, p. 88.

[8] Newton, *Opticks*, pp. 405-6, cited also in this book on chapter II. On this issue see also: McGuire and Rattansi, "The Pipes of Pan," and Rattansi, "Newton and the Wisdom of the Ancients," pp. 198-9.

[9] Dobbs, *Janus Faces*, p. 72.

philosophy capable of imitating God's creation as well as replenishing the corrupted stage of the material world. Furthermore, as Karin Figala writes, it "was especially in alchemy that the great savant, beginning with the outermost, thinnest layer of the body, attempted to advance towards the innermost, most concentrated and noble invisible seed-like core, in order to observe its maturation and growth at different stages."[10] This imagined procedure of breaking into the secrets of the material design and imitating its stages of birth, growth and death underlies Newton's insistence that the alchemical process was "quite explicitly parallel to God's creative activity at the beginning of time."[11] On this he writes:

> And just as all things were created from one Chaos by the design of one God, so in our art all things ... are born from this one thing, which is our Chaos, by the design of the Artificer and the skilful adaptation of things. And the generation of this is similar to the human, truly from a father and mother.[12]

Dobbs points out this belief was not as strange as it may sound to the modern ear, since it "was a common assumption in the 17th century that the physical state of our world paralleled the moral history of mankind."[13] Indeed, Newton and Whiston shared the view that:

> wise men would rather set themselves carefully to compare Nature with Scripture, and make a free Enquiry into certain *Phaenomena* of the one, and the genuine Sense of the other; which if Expositors would do, 'twere not hard to demonstrate in several such cases, that the latter is so far from opposing the truths deducible from the former.... That 'tis in the greatest harmony therewith.[14]

Whiston was much more outspoken regarding the paralleled history of nature and humanity than Newton. For example, he argued that a comet caused

---

[10] Karin Figala, "Newton as Alchemist," pp. 102-37.

[11] B.J.T. Dobbs, "Newton's *Commentary* on the *Emerald Tablet* of Hermes Trismegistus: Its Scientific and Theological Significance," I. Merkel and A.G. Debus, (eds.), *Hermeticism and The Renaissance* (Folger Books, 1988), pp. 183-91, 185; Figala, "Newton as Alchemist," pp. 126-8.

[12] Newton, Keynes MS. 28, fol. 2*r*, *v*, in Dobbs, "Newton's *Commentary* on the *Emerald Tablet*," p. 185.

[13] Dobbs, *Janus Faces,* p. 232.

[14] William Whiston, *A New Theory of the Earth* (London, 1696), p. 64, cited also in Force, *Whiston,* p. 41.

the end of paradise changing the earth's shape from a slightly eccentric to a concentric one; and the impact tilted the earth's axis, producing seasons, winds, tides, and diurnal rotation. Further more, Whiston suggested that the natural and mechanical conjunction of the course of the comet with the precise moment when it was necessary to punish excessive wickedness is:

> the secret of the divine providence in the world.... He who has created all things... At once looks at the intire Train of future causes, actions and events, and sees at what periods, and in what manner 'twill be necessary and expedient to bring about any changes, bestow any Mercies, or inflict any Punishments on the World. ... So that when the Universal Course of Nature, with all the Powers and Effects thereof, were at first deriv'd from, and are continually upheld by God; and when nothing falls out otherwise, or at any other time, than was determin'd by divine Appointment in the Primitive Formation of the Universe: To assign *Physical* and *Mechanical* causes for the Deluge, or such mighty judgments of God upon the Wicked, is so far from taking away the Divine Providence therein, that it supposes and demonstrates its Interest in a more Noble, Wise and Divine manner than the bringing in always a miraculous Power wou'd do.[15]

In such a context it is less surprising to discover that Newton explored the secrets of alchemy. Following the alchemical tradition, he believed that the world was created out of a state of putrefaction, of material chaos, through God's divine light. After creation, an inevitable Fall brought about deterioration, corruption and expansion, followed by a new restoration of truth instilled by God through a messenger soul. Then once again comes a fall, a new restoration, and so on, until the whole cycle reaches a destructive end as foretold by the prophets. Dobbs points out that Newton's Arian beliefs intermingled with his alchemical studies, especially his belief in the role of Christ as God's spiritual agent in the world. In his alchemical and theological studies prior the *Principia*, Newton believed that "brought into being by the Father before all worlds and all ages, the Son in his turn became the framer of the cosmos, the agent of God's creativity." Both Father and Son had a union in dominion but not in substance. "The will of God for creation and guidance was worked through the agency of the Son, the creative Word or Logos. Thus Christ was the viceroy, the spiritual being that acts as God's agent in the world." Newton identified Christ with the Word and argued that Christ was with God before his incarnation, even in the beginning, and that "he was God's

---

[15] Whiston, *Memoirs of the Life and Writings of Mr. William Whiston. Containing Memoirs of Several of his Friends also*,2nd ed., (2vols., London, 1753) vol. 1, pp. 250-1, cited also in Force, Whiston, p. 47.

active agent throughout time, speaking to Adam and Moses by the name of God."[16] According to Newton, God told the Son to walk in the Garden of Eden with Adam, to appear as an angel to Abraham, to fight with Jacob, to give the prophecies to the prophets, and to preach the pure religion to Noah, the Jews, and the Gentiles.[17] Or, in his own terms:

> Christ was with God before his incarnation, even in the beginning when God made the heavens & earth. For Christ himself declared as much when he said to his Father: *Glorify me with the glory which I had with thee before the world began.* It signifies that he being then with God, it was he to whom God said Let us make man, & That it was he who appeared to Adam in paradise by name of God, and to the Patriarchs & to Moses by the same name: For the father is the invisible God whom no eye hath seen nor can see. It signifies that he was one of the three Angels who appeared to Abraham & of whom it is said Jehova rained upon Sodom & Gomorrah brimstone & fire from Jehova out of heaven: ... the name Jehova is given to none but the God of Israel. It signifies he is the God who wrestled with Jacob & to whom Jacob erected an altar Gen. 35.1, 12, and the Angel who appeared to Moses in the bush by the name of the God of his fathers Abraham Isaac & Jacob, (Exod. 3) & was with Moses in the wilderness & spake to him on mount Sinai giving lively oracles to the people & to whom the people were disobedient thrusting him from them & worshipping the Calf (Act. 7.38, 39, 40). It signifies that the he was the Angel of God's presence of whom God said to Israel *Behold I send an Angel before thee to keep thee in the way & to bring thee into the place which I have prepared: beware of him & obey his voice, provoke him not: for he will not pardon your transgressions: for my name is in him.* Exod. 23.20.21....
>
> Christ is therefore called the Word to signify that before his incarnation he was the Oracle & mouth of God the Angel by whom God gave the law on mount Sinai & commanded Israel & whose voice to be obeyed: And also to signify that in his mortal body he was the Prophet predicted by Moses: & that after his resurrection he was the faithful and true witness, whose testimony was the spirit of prophecy & who shall come

---

[16] Dobbs, *Janus Faces*, p. 82. Dobbs adds on this that even though Newton's God is exceedingly transcendent, God never loses touch with his creation, for He always has the Christ transmitting his will into action in the world. Ibid, p. 83.

[17] On this issue see: James E. Force, "The Newtonians and Deism," *Essays on the Context, Nature, and Influence of Isaac Newton's Theology*, pp. 53-66.

> to destroy the wicked with the breath of his mouth as with a two edged sword, & to judge the quick & the dead.[18]

On similar lines, Dobbs adds that Newton's association of the vegetable spirit of alchemy with the cosmic governance of the Arian Logos has led Newton to assume that Hermes, himself a personification of the operative alchemical spirit, was a pagan type of Christ.[19] Yet, according to Newton, only a select few, a remnant, could grasp the secret behind Christ as the Logos.[20] The "world loves to be deceived, they will not understand," such spiritual matters, writes Newton, since "they never consider equally, but are wholly led by prejudice, interest, the praise of men, & authority of the church they live in." Unfortunately, there "are but a few that seek to understand the religion they profess."[21] Those select few wise men, who have become a practical demonstration of the religion they profess, remind the corrupted people of the true spiritual loving relationship souls have with God and amongst themselves:

> The moral part of all religion is comprehended in these two precepts: Thou shalt love the Lord thy God with all thy heart & with all thy soul & with all thy mind. This is the first & great commandment & the next is like unto it, Thou shall love thy neighbour as thy self. Upon these two hang all the law & the Prophets. These are the laws of Nature, the true natural religion.[22]

Nonetheless, as time goes on, the inevitable fall and degradation towards moral and material corruption recur. People's intellects are pulled towards materialistic ends and become unable to grasp God's simple truth. They begin to worship idols at the prompting of their intellects' degraded status. Their natural philosophy and morality also deteriorate. Even when corruption reaches such an advanced stage, alchemical knowledge can still transform the most degraded state of matter into its most divine golden nature. The alchemical process transforms the deteriorated state of matter into its highest and original divine form from which a new natural cycle of elevation and degradation can start afresh. The alchemical process first transforms the stone into a black chaos, a *materia prima.* At that stage of putrefaction a new

---

[18] Newton, Yahuda MS. 15, Ch.2, ff. 96-7, cited in Castillejo, *Expanding Force*, p. 77.

[19] Dobbs, *Janus Faces*, p. 150.

[20] Indeed this secret is puzzling, more so when we add to it Newton's God of dominion and his interpretation to the Apocalypse. See the discussion in the next chapter.

[21] Newton, "*A Treatise on the Apocalypse*," Yahuda, MS. 1 f.5r.

[22] Newton, MS. Bodmer,Ch.7, "Chap. Of the Christian Religion and its corruption in Morals," p.2r, also in: Goldish, *Judaism*, p. 41.

spiritual seed is planted with the assistance of united opposites and as the sample matures it becomes a more divine form of matter. This form will once again become corrupted, putrefied, and transformed into a higher divine form:

> [O]nce an aggregate has been formed, the agent's (mercury) first act is to putrefy the aggregate and confound it to chaos. Then it proceeds to generation.... It accommodates itself to every nature. From metalic semen it generates gold, from human semen men, etc. and it puts on various forms according to the nature of the subject.[23]

> As the world was formed of a chaos or materia prima which was void and without form: so in our work the stone is made of chaos or materia prima which by putrefaction is void without form and black and dark.[24]

> Inferior and Superior, fixed and volatile, sulfur and quicksilver have a similar nature and are one thing, like man and wife. For they differ one from another by the degree of digestion and maturity. Sulfur is mature quicksilver and quicksilver is immature sulfur; and on account of this unity they unite like male and female, and they act on each other, and through that action they are mutually transmuted into each other and procreate a more noble offspring to accomplish the miracles of this one thing.[25]

The alchemical process is actually an imitation of the ways Nature works in the generation of living beings. All processes in the living world are cyclical:

> For Nature is a perpetual circulatory worker, generating fluids out of solids, and solid out of fluids, fixed things out of volatile, & volatile out of fixed, subtile out of gross, & gross out of subtile.[26]

---

[23] Newton, Keynes MS. 12A (I, n.29), ff. Iv-2r, also in Dobbs, *Janus Faces*, p. 25.

[24] Newton, citation from "Out of La Lumiere," Dobbs, *Janus Faces*, p. 174.

[25] Newton, Keynes MS. 28 (I, n.35) f.6r,v, also in Dobbs, *Janus Faces*, p. 70. On a more detailed account on the alchemical definition of sulphur as mature mercury and mercury as immature sulphur, see: Figala, "Newton as Alchemist," pp. 125-6.

[26] Newton, *The Correspondence*, vol 1, pp. 365-6.

To simplify the logic behind the alchemical cycle I would like to suggest the following abstraction: three tendencies are manifest in alchemy and in the cycle of nature - a tendency to destroy the old corrupted and deteriorated stage, a tendency to create, and a tendency to sustain and preserve the begotten form. The result of the tendency to destroy is putrefaction, which is a state of materia prima – a black chaos, which becomes a potential womb for all forms. Putrefaction is a state of disorganization in nature where all the parts of matter become similar to each other. This stage is actually the natural tendency of passive matter to disintegrate. At this disorganized stage passive matter is spread out evenly in space and is the best potential chaos for a new material organization[27] around an emanating informative spirit.[28] Or, in Newtonian terms:

> Nothing can be changed from what it is without putrefaction.
> No putrefaction can be without alienating the thing putrefied from what it was.
> Nothing can be generated or nourished (but of putrefied matter).
> All putrefied matter is capable of having something generated out of it and in motion toward it.
> Her action is to blend and confound mixtures into a putrefied chaos. Then they are fitted for new generation or nourishment.
> All things are corruptible.
> All things are generable.
> Putrefaction is the reduction of a thing from that maturity and specification it had attained by generation.[29]

> The general method of operation of this [vital] agent [mercurial spirit] is the same in all things; that is, it is excited to action by a gentle heat, but driven away by a great one, and when it is introduced into a mass of substances its first action is to putrefy

---

[27] This chaotic state reminds one of the modern concept of entropy, a material symmetrical state with no order.

[28] The children's game called Lego is a good exemplification of the alchemical process of an organized material design first putrefying and organizing itself again according to the instructions of a spiritual seed. A child builds a house with the smaller Lego blocks and then wants to build a new structure, say a farm. The design of the farm is drawn on paper, specifying also the colors and the size of each building block. Will she build the new design from the structure of the old building? Or will she first break down the old house to the smallest building blocks (process of putrefaction), organize them according to size and color and only then start building the farm according to the instructions (the spiritual seed written as a code)?

[29] Newton, Dibner collection MSS 1031B f.5r.4, Dobbs, *Janus Faces*, p. 56.

> and confound into chaos; then it proceeds to generation.[30]

This tendency to destroy the previous form operates in nature whenever nature reaches a corruptible stage:

> Hence also it may be, that the Parts of Animals and Vegetables preserve their several Forms, and assimilate their Nourishment; the soft and moist nourishment easily changing its Texture by a gentle Heat and Motion, till it becomes like the dense, hard, dry, and durable Earth in the Center of each Particle. But when the Nourishment grows unfit to be assimilated, or the Central Earth growth too feeble to be assimilated, or the Central Earth grows too feeble to assimilate it, the Motion ends in Confusion, Putrefaction, and Death.[31]

A similar pattern occurs in Newton's description of world history. All the important prophet souls appeared at a stage of corruption in the history of humanity, which was followed by destruction. Social life became too "feeble to be assimilated" into virtue, thus a new divine light, a revelation of truth, was necessary for recovery:

> So then the first religion was the most rational of all others till the nations corrupted it. For there is no way {(wthout revelation)} to come to the knowledge of a Deity but by the frame of Nature.[32]

As I understand Newton's systematic and analogical way of organizing knowledge in different realms, Noah's case, as well as that of other messenger souls, presents a cyclical pattern similar to that of alchemy. The new religion Noah established is an illustration of the instillation of spiritual goodness planted as a new concentrated seed in the midst of social destruction. In Noah's days civilization became so corrupted that God considered destroying it. The revelation of truth occurred to Noah only a short time before the flood. As long as people were content with their idolatrous behavior no one listened to him. Only after the flood could this new concentrated spiritual seed sprout since the flood created a stage of social putrefaction. Similarly, in alchemy the tendency to destroy transforms the corrupted stage into a putrefied chaos in which a new seed of truth can be planted. Indeed, all of the new

---

[30] Newton, Keynes MS. 12, ff.1v-2, cited also in Westfall, *Never at Rest*, pp. 304-5.

[31] Newton, *Opticks*, p. 387.

[32] Newton, MS. Yahuda 41, p.7r,

restorations of Noah's religion by messenger souls may be seen as a new purified and concentrated seed of spiritual truth planted in the midst of social chaos. The case of Moses is also exemplary. The Israelites were able to hear and accept Moses' spiritual message only after they had gone through slavery and great suffering, thus reaching a very severe form of human putrefaction.

The second tendency of nature to create occurs when the seed is planted in the black putrefied chaos. Maturation of a spiritual seed will not succeed if putrefaction is not present. Putrefaction and social chaos are crucial elements for the beginning of the alchemical cycle and for the revelation of God's true religion. A seed cannot sprout in an over-organized material system since the previous order limits the possibilities of the new material organization to grow in accordance with the spiritual design concentrated as a code in a seed.[33] In Newton's early paper on the nature of light, which he sent to Oldenburg in December 1675, he was very clear about this tendency to generate forms inherent in nature out of concentration:

> Perhaps the whole frame of Nature {may be nothing but aether condensed by a fermental principle} ... and after condensation wrought into various forms, at first by the immediate hand of the Creator, and ever since by the power of Nature, wch by virtue of the command Increase and Multiply, became a complete Imitator of the copies sett her by the Protoplast.[34]

In the midst of a state of primeval chaos (putrefaction), God, as the Protoplast, imprints his divine forms upon matter, later to "Increase and Multiply" by nature according to his biblical command. McGuire assumes this text expresses Newton belief that "by a sovereign decree God brought the world into existence. Once established, however, that order maintains itself by 'the power of Nature.' Which in turn enacts the Divine decrees by generating phenomena according to the protoplast or archetype. These copies are therefore physical imitations, manifesting physical laws."[35] Dobbs suggests another interpretation of the meaning of "Protoplast" and "increase and multiply." The word protoplast, she thinks, has two meanings. The first meaning is to be understood as the models, the exemplars, the originals of all created things (first formed), made by God. Nature cannot form protoplasts. Only God can do

---

[33] Put more simply, if one wants to insert a newer version of a computer program, one needs to delete the old one. The deleting of the old program assures that the infrastructure of the computer is once again in a stage of "putrefaction," so to speak; the logical circuits of the computer are not controlled by software instructions, which would prevent the functioning of the newer version of the program.

[34] Newton, *The Correspondence*, vol. 1, p. 364.

[35] McGuire, "Newton's Invisible Realm," p. 192.

that, though nature can nourish them.[36] The second, less common, meaning is the first former or creator. Dobbs argues that this second meaning, as first agent in the creative process, reflects Newton's shift to the Arian heresy. In it we find his belief in a transcendent Arian Deity who interacts with the world through the active agency of the Logos, the Arian Christ. The Arian Christ prepared and formed the cosmos and continues to govern it according to God's will, constituting God's "immediate hand" in all the contacts between God and man reported in the Bible. "Within this Arian context then the 'Protoplast' that has fashioned the copies for nature to imitate must be interpreted not as the supreme Deity but rather as the Christ acting in his capacity as cosmological agent."[37]

In the queries on the nature of light written much later in his life, Newton gave nature the role of preserving the first meaning of the protoplast as the model, the exemplars. He writes, "Nature is very consonant and conformable to herself."[38] To achieve the copying of the divine forms it is necessary that nature employ its third tendency - to preserve – that is, be "very consonant and conformable to herself." The power of nature to maintain the original divine order given to it by the Creator, either directly or through Christ, the Logos, is the third tendency, that of sustaining order. But before going on to this third tendency to preserve forms, more discussion of the second tendency, to create, will be necessary. Newton identified this second tendency with active agents such as gravity, fermentation, and the cohesion of bodies:

> It seems further, that these Particles have not only *Vis inertiae*, accompanied with such passive Laws of Motion as naturally result from that Force, but also that they are moved by certain active Principles, such is that of Gravity, and that which causes Fermentation, and the Cohesion of Bodies. These Principles I consider not as occult Qualities, supposed to result from the specifick Form of Things, but as the general Laws of Nature, by which the Things themselves are form'd: Their Truth appearing to us from Phenomena, though their Causes not yet discovered.[39]

This text shows Newton's view of the divine secrets behind creation and the "increase and multiply" of the protoplast (in its first meaning of a model and exemplar). Gravity, fermentation, and the cohesion of bodies create and generate the "specific form of things." As Dobbs shows, at a certain period in Newton's life he held that gravity was a cosmic music that created

---

[36] Dobbs, *Janus Faces*, p. 107.

[37] Dobbs, *Janus Faces*, p. 110.

[38] Newton, *Opticks*, p. 376.

[39] Newton, *Opticks*, p. 401.

mathematical relationships amongst celestial bodies in the universe. Gravity was the only kind of spirit that could penetrate the center of bodies without causing retardation. The different constellations of the heavenly stars are analogous to musical instruments played by the Creator, and gravity is the notes through which God plays the harmony of the cosmos. At creation God imprinted the model or exemplar of these pristine mathematical relationships upon matter and from then on his divine music plays continuously. Newton maintained that "the *music of the spheres* and the *pipes of Pan*," as Rattansi puts it, is music to which "the whole of creation was said to dance."[40] God constantly plays this heavenly music:

> What is it, by means of wch, bodies act on one another at a distance. And to what Agent did the Ancients attribute the gravity of their atoms and what did they mean by calling God a harmony and comparing him & matter (the corporeal part of the Universe) to the God Pan and his Pipe.[41]

> For Pythagoras, as Macrobius avows, stretched the intestines of sheep or the sinews of oxen by attaching various weights, and from this learned the ratio of the celestial harmony… But the proportion discovered by these experiments, on the evidence of Macrobius, he applied to the heavens and consequently comparing those weights with the weights of the Planets, and the length of the strings with the distances of the Planets, he understood by means of the harmony of the heavens that the weights of the Planets towards the Sun were reciprocally as the squares of their distances.[42]

After the creation of the protoplast, without further ado God can sustain and play the heavenly music. He does so merely through the presence of his eternal equable flowing duration purposefully designed in accordance with the material constitution of the solar system. To reiterate: all material atoms are mathematical quantities (similar to the mathematical nature of the absolutes), immutable (as space) and have *vis insita* (flow equably as absolute time). Moreover, the sun as the fire at the heart of the world represents God's state of detached purity radiating eternally, acting without being influenced. This is the deeper meaning of the sacrificial fire and its relationship to the divine music. Indeed, according to Dobbs, in the middle period of his life,

---

[40] Rattansi, "Newton and the wisdom of the ancients," p. 198.

[41] Newton, Draft of Query 27 of *Opticks*, cited in McGuire and Rattansi, "Newton and the 'Pipes of Pan,'" p.108.

[42] Newton, "Classical Scholia," written for Proposition VIII, cited also in: James, *The Music of the Spheres*, p. 165.

Newton concluded that "the sun was at the center of the penetrating gravitational attraction that held his heliocentric world together as it was the source and celestial counterpart of the fermental virtue that animated, shaped and governed the realm of passive matter."[43] Both gravity and fermentation are active divine forces, which act and organize matter without being influenced by it.

As I understand Newton's obscure writings on this issue, gravity and fermentation function analogously to the second and third order of computer languages written on the hardware as a mathematical and logical design specifically suited to the material realm, so that it can connect with the eternal flow of the absolutes in the simplest possible way to create any possible chosen order. In God's chosen order for our world, the sun, for example, is the most essential equipment for the running of the whole order of the design, it functions like the heart as a source of energy to the whole system. Close to Newton's views on the role of the gravity was the Hermetic tradition of the sun as the channel by which creative energy passed from the super-celestial realm to the elemental or terrestrial realm. It was the fecundity of the sun, its action as demi-urge in generating life and in serving to form, conserve, and nourish all the infinite variety of things that attracted Newton.[44] Indeed, without the sun life could not exist. The sun in this tradition is like a huge power station installed in the world computer through which the divine current enters and runs the program. If this power station fails the whole materialized order of the design will collapse since not enough current will enter the computer for its regular functioning. However, the design itself, like the circuitry of the CPU (central processing unit of a computer), the brain of the computer, is not influenced by the fall of the power source. The brain functioning of the computer of the worldly design is independent of the energy activating it. The design is not conditioned by the current (absolute time) and in principle it can exist as a non-realized spiritual world. The input of the eternal flux of energy transforms the wisdom of this spiritual universe into a material world, from which human beings too are constructed, so that they can come to perceive the order of the design.[45]

---

[43] Dobbs, *Janus Faces*, p. 166.

[44] Dobbs, *Janus Faces*, pp. 159-60, for precise reference on the Hermetic tradition, see there in n. 116, p. 159.

[45] Another explanatory analogy that comes to mind is the following: Absolute space has two functions. The first function is to be the root of all places, or the place of all places (M*akom)*. As such absolute space must contain the divine design within its fabrics, because it is the root of all presence, so to speak. This function enables God's equable flowing energy (called absolute time) to be transferred as a pipeline into this pre-designed *Makom* whilst governing the intensity of the divine presence of all created material beings according to the design. Nowadays physicists believe that there is huge latent energy in the vacuum, and that the fluctuation of this energy can form a whole universe. That is one explanation to the recent observed phenomena of the accelerated expansion of the universe. The vacuum acts as a reservoir of God's energy. The analogy of the computer is thus not complete since in computers the screen does not supply the energy for the CPU as the vacuum. This brings us to the

As I see it, for Newton souls and matter have two different kinds of relationships with God - one pure, simple, and direct, the other involving different degrees of impurity which disconnect them from the eternal Source of energy and create complexity and corruption. Matter is pure as long as it is completely controlled by the original design given to it by the Creator. But when humanity falls into idolatry a similar thing happens to matter. Instead of the harmonious and loving exchange amongst souls and the supreme soul, attuned to the heavenly music represented by the worship of the sacrificial fire, humans as well as heavenly bodies are no longer synchronized with God's original music. People become arrogant and corrupt and impose bondage and destruction on each other. Rather than being attuned to the one eternal current emanating unconditionally from the source, they begin to be influenced by one another. They add redundant complexities to the simple design, obstructing the daily unconditional sustenance entering it. In this sense idolatry promotes corruption and disruption in the original divine design and turns matter into a less efficient divine conductor.

The possibility of becoming disorganized is inherent in the passive laws of matter since atoms are imbued with an inert passive force. These laws were designed for the daily sustenance given to material bodies by God, and for creating an aggressive reaction of bodies towards any intrusion or change in the original divine state God gave them at creation. *Vis insita* of bodies has this inherent dual aspect, which is closely connected to the third tendency inherent in nature to sustain forms. The tendency to sustain forms once created and to preserve them in their original state around central forces is also built into the nature of passive matter:

> [*Vis Insita*] is ever proportional to the body whose force it is, and differs nothing from the inactivity of mass, but in our manner of conceiving it. A body, from the inactivity of matter, is not without difficulty put out of its state of rest or motion.[46]

Vis insita is a divine state of atoms that flows equably along absolute time, yet it alone cannot explain the composing of atoms into divine forms. Once God's protoplast is imprinted upon bodies through active central divine forces, such as gravity, this passive force is charged only with sustaining the

---

second function of absolute space. Here space serves as a screen that reveals the hidden design to observers, as a computer screen that displays the design. The computer runs separately from the screen, yet as long as it is not connected to the screen nothing of the historical unfolding of the order of the design will be seen. The screen itself does not perceive what is screened on it, but without it nothing can be sensed. This might explain why Newton calls absolute space God's sensorium; without its existence the design of creation would never become visible to the senses since the order of the design would not be displayed.

[46] Newton, *Principia*, Book I, definition III, pp. 9-10.

state of rest or constant motion of the atoms composing the bodies in accordance with the original divine plan. "For a body maintains every new state it acquires," writes Newton, "by its *vis inertiae* only."[47] Yet this sustenance is very fragile, since "motion is much more apt to be lost than got, and is always on the Decay."[48] To my mind the reason for this natural decay is inherent in the dual reaction of *vis insita* to any change in the inertial state of bodies:

> The body exerts *vis insita* only, when another force, impressed upon it, endeavours to change its condition; the exercise of this force may be considered both as resistance and impulse; it is resistance, in so far as the body, for maintaining its present state, withstands the force impressed; it is impulse, in so far as the body, by not giving way to the impressed force of another, endeavors to change the state of that other.[49]

This dual aspect of resistance and impulse is even more explicit in Newton's third law of motion of action and reaction. On this law he writes that "whatever draws or presses another is as much drawn or pressed by that other."[50] Thus at the moment an external force is impressed upon a body (gravity and fermentation in their original forms are not considered external in this way since they are God's direct intervention) the body protects itself by resisting the change; yet at the same time it attacks the intruder and brings about change in the equable flow of the momentarily intruding force as well as in its own state. Thus in any interaction of this sort the purity and order of the original protoplast is necessarily reduced, since the aggressive exchange has removed momentarily the body from its peaceful flow with God's absolute time. A new state of inertia is acquired when the momentary intruding force stops, yet the new acquired stage which flows once again along God's equable flow is less attuned to the music of God since it was removed from its original ordered state due to its aggressive stance and now it mistakenly identifies its new state as the one that needs to be protected.

More precisely, God is the only being that can act upon matter without being reacted upon. Only the actions of God himself upon matter do not create a chain of action and reaction, or material bondage, so to speak, since they are a prior spiritual design; thus as long as the body is completely attuned to God's music there is no interruption of its original state. Motions that are governed directly by God are called absolute motions, whereas all relative motions are due to interactions that are disruptive to the original plan; that is, a chain of

---

[47] Newton, *Principia*, Book I, definition IV, p. 10.

[48] Newton, *Opticks*, p. 398.

[49] Newton, *Principia*, Book I, definition III, p. 10.

[50] Newton, *Principia*, Book I, Law III, p. 19.

action and reaction is generated. Any chain of action and reaction triggered by relative motions separates matter from God's original music. This process of decline is even clearer in the history of humanity.

As long as humans worship according to the pristine religion they exchange only love amongst themselves and God. This is a phase of absolute worship expressed through loving relationships amongst people and God. Yet once people's intellects are no longer attuned only to God's simple creed and they become materialistic, they gradually become influenced by one another and by matter and a chain of bondage and corruption extends and expands. Purity is lost in this "relative" exchange amongst humans and bodies, since only the direct and absolute connection with God establishes truth and love in society. Our relation to God is the following:

> God made the world and governs it invisibly, and hath commanded us to love, honour and worship him and no other God but him and do it without making any image of him, and not to name him idly and without reverence, and *to honour our parents, masters and governors,* and love our neighbors as ourselves, and to be temperate, moderate, just and peaceable, and to be merciful even to brute beasts.[51]

When human beings do not follow these commandments corruption sets in. We find this idea when Newton determinedly defines idolatry as:

> breach of the first & greatest commandment. It is giving to idols the love honour & worship wch is due-to the true God alone. It is forsaking the true God to commit whoredome with other lovers.[52]

Idolatry as a form of "whoredom" bonds people in a "relative" relationship with matter and with their ancestors, and no longer admits a direct loving connection with God. Any form of impurity and corruption is established when souls instead of being attuned to the original religion of Noah and to God's heavenly absolute music become lost and trapped in material desires. When human beings begin to worship creation (in the form of an idol or another human being), and not the Creator, they are no longer fully attuned to the divine music. They start interacting in an action-reaction mechanism, accelerating "relative" motion in the world. The decay and corruption of creation and humanity become inevitable. The only true absolute loving and active source of energy is God. Idolatry is a corrupted interaction, since it is not

[51] Newton, "Irenicum," Keynes MS. 3, cited also in Force, "Newton's God of Dominion," p. 94.

[52] Newton, Keynes MS. 3 f. 14r, also cited in Force, "Newton, the Lord God of Israel," 143.

attuned to the source. Throughout history, God sustains the original forms of creation through the music of gravity. Idolatry and material interactions oblige him to replenish creation whenever such interactions corrupt the original design to such a degree that society and the whole solar system are on the verge of collapse.

The cycle of destruction, putrefaction, generation, sustenance, corruption, and once again destruction, is inevitable. The alchemist's work is to assure that any original form that has reached a severe form of corruption will be destroyed into putrefaction and then transformed into a new divine form through the fire of sublimation. This is the true meaning of the sacrificial fire. It is the fire that burns and transforms corruption into putrefaction whilst planting a new divine seed to start a new cycle. This cycle is described in Newton's notes from *La Lumiere*:

> But the seed is unprofitable unless it rot & become black, for corruption always proceeds generation & we must make black before we can whiten.[53]

Both nature and human history follow cyclical processes. "The changing of bodies into light," writes Newton in the queries, "and light into bodies, is very comfortable to the course of Nature, which seems delighted with transmutations."[54] God has created the protoplast of all nature and society. First the true absolute worship of God and the divine and loving relationship amongst human beings were established, then corruption in the form of idolatry crept in, followed by the expansion of society and material bondage. Nonetheless, whenever the spread of idolatry attains a corruptive state, God, the light, the eternal flame, intervenes and replenishes creation. The intervention comes in a form of semi-destruction followed by putrefaction and the generation of another form of life.

Where and how does the method of fluxions fit into this cycle of generation and creation? This method is crucial for the understanding of the mechanism of the daily sustenance and the regaining of human control over nature. The method of fluxions, which calculates the rate of fluxion of quantities from the perspective of an absolute divine equable flow, enables men to regain control over a creation that has lost its true connection with the Creator. The mathematical clarity of the method enabled Newton, so to speak, to regain control over relative motions and recognize the absolute heavenly music of gravity known to the ancients:

> To some such laws [of motion arising from life] the ancient Philosophers seem to have alluded when they called God

[53] Newton's notes from *La Lumiere*, in Dobbs, *Janus Faces*, p. 281.

[54] Newton, *Opticks*, p. 374.

> Harmony and signified his actuating matter harmonically by the God Pan's playing upon a Pipe and attributing musick to the spheres made the distances and motions of the heavenly bodies to be harmonical, and represented the Planets by the seven strings of Apollo's Harp.[55]

Absolute divine time, which is beyond the corporeal stage, is eternal and non-destructible, always flowing equably and peacefully. In contrast, the relative time of the world of action and reaction, where the drama of creation takes place, is cyclical. It is the result of the union of two divine opposing forces acting on all material bodies, one flowing passively, equably and unconditionally; the other, as gravity and fermentation, momentary in respect to eternity, dynamic, and sometimes disruptive. Figala associates these opposing forces with Newton's cosmic time and alchemical work as follows: in "celestial mechanics the centrifugal force (*vis insita, vis inertiae*, force of inactivity) would mean the fixing, formative, sulphuric; and the centripetal force (dynamic force, *vis impressa*) would mean the volatile mercurial. Only the union of the two opposing single forces results in the ideal closed circular movement generally assumed by Newton to be the 'Perpetual Worker of Nature.'"[56]

As with everything else, Newton's belief in the cyclical cosmogony and the search for the *prisca sapientia* is totally opposed to Leibniz's understanding of the role of history and progress. On these issues Leibniz writes:

> In general, one may say that though afflictions are temporary evils, they are good in effect, for they are short cuts to greater perfection.... To the objection which could be offered [that destruction and decline lead to a better result], moreover, that if this were so, the world should long since have become a paradise, there is an answer near at hand. Although many substances have already attained great perfection, yet because of the infinite divisibility of the continuum, there always remain in the abyss of things parts that are still asleep. These are to be aroused and developed into something greater and better and in a word, to a better culture. And hence progress never comes to an end.[57]

---

[55] Newton, a draft for the Scholium intended for Proposition IX, cited in McGuire and Rattansi, "Newton and the 'Pipes of Pan,'" p.118.

[56] Figala, "Newton as Alchemist," p. 128.

[57] Leibniz, "On the Radical Origination of Things," *G.W. Leibniz: Philosophical Papers and Letters*, pp. 490-1.

For Leibniz, time as an order of succession in the best possible world, has no disastrous end and is not cyclical; on the contrary, it is progressive. Time is an infinite series of predestined perceptual monadic states internal to individual monads, each expressing the progression from its own particular point of view. The expression of truth and goodness unfolds as monadic states develop. One cannot find the expression of truth in the remote perceptions of the past. Newton's system, in contrast, expresses an opposite historical narrative. The remote past is a much clearer and simpler expression of truth; the future involves complexity, confusion and destruction. To decipher the simplicity of the ancient wisdom one needs a proper historical understanding of the work of providence as expressed in prophecy. Alchemy, though a very old science, is nonetheless not the most pristine of sciences. It already carries a minor corrupt interpretation of the scientific truth. The pristine expression of the scientific truth of the worldly design, as well as the historical laws governing human corruption, were contained, Newton thought in the ancient Jewish ceremonies around the Tabernacle and the two Temples. In what follows, I shall attempt to set out this true ancient science according to Newton's cryptic and scattered writings and notes.

CHAPTER IX

THE TABERNACLE AND THE TWO JEWISH TEMPLES

> "In the Apocalypse the world natural is represented by the Temple of Jerusalem & the parts of this world by the analogous parts of the Temple." (Newton, Keynes MS. 5 vr. 9.)

Newton interpreted the prophecies of Daniel and John as possessing the ultimate secret of the history of the world, condensed in a series of visions:

> The predictions of things to come relate to the state of the church in all ages: and amongst the old prophets, Daniel is most distinct in order of time, and easiest to be understood: and therefore in those things which relate to the last times, he must be made the key to the rest.[1]

> The prophecies of Daniel are all of them related to one another, as if they were but several parts of one general prophecy at several times. The first is the easiest to be understood, and every following prophecy adds something new to the former.[2]

Newton's approach to Daniel's and John's prophecies shows his attitude to the words of prophecy. On these issues he writes:

> In the infancy of the nation of Israel, when God had given them a law, and made a covenant with them to be their God if they would keep his commandments, he sent prophets to reclaim them, as often they revolted to the worship of other gods: and upon their returning to him, they sometimes renewed the covenant which they had broken. These prophets he continued to send till the days of Ezra: but after the prophecies were read in the Synagogues, those prophecies were not sufficient. ... At length when a new treaty was to be preached to the Gentiles,

---

[1] Newton, *Observations upon the Prophecies of Daniel and the Apocalypse of St. John*, Samuel Horsley (ed.), *Isaaci Newtoni Opera Quae Exstant Omnia* (5 vol's; London, 1779-1785), vol. 5, part I, #xvii, p. 305.

[2] Newton, *Observations*, part I, chapter 3, p. 311.

namely, "that Jesus was the Christ," God sent new prophets and teachers: but after their writings were also received and read in the Synagogue of the Christians, prophecy ceased a second time. We have Moses, the prophets, and apostles, and the words of Christ himself, and if we will not hear them, we shall be more inexcusable than the Jews.... The giving ear to the prophets is a fundamental character of the true church. For God has so ordered the prophecies, that in the latter days "the wise may understand but the wicked shall do wickedly, and none of the wicked shall understand." (Dan. xii. 9, 10.)

The authority of emperors, kings, and princes, is human. The authority of councils, synods, bishops, and presbyters, is human. The authority of the prophets is divine, and comprehends the sum of religion, reckoning Moses and the apostles among the prophets... Their writings contain the covenant between God and his people, with instructions for keeping this covenant; and predictions to things to come.[3]

I. CURRENT RESEARCH

As Castillejo and Goldish have observed, to understand Newton's ideas of these prophecies and their functioning we need to see the historical connection he made between the Tabernacle, the first Temple and the second Temple. Both the physical form of these places and the rites conducted around them were extremely important to him. The structure of the Tabernacle and the Temples encoded the laws of nature, whereas the ceremonies prefigured historical events. The physical form carried the image of the sun-centered universe, and was also the site of the prophetic events described in the Book of Revelation. To decipher the prophecies Newton meticulously reconstructed the Temples of Solomon and Ezekiel, because every detail was a prefiguration.

The reconstructed shape of the Temple appears in the published *Chronology* (Chapter V) and in the manuscript *Prolegomena ad Lexici Prophetici* (Babson MS. 434). Both texts give figures of Temple measurements.[4] Also important is the posthumously published *Dissertation upon the Sacred Cubit of the Jews*, which discusses the length of the sacred cubit used in building the Temple in relation to the cubits of other ancient

---

[3] Newton, *Observations*, Part I, # xvi, pp. 304-5, my emphasis.

[4] Isaac Newton, *Chronology of Ancient Kingdoms Amended* (London, 1728), chapter 5. Ciriaca Morano has prepared a scholarly edition of *Prolegomena ad Lexici Prophetici Partem Secundum in quibis Agitur DeFforma Santuarii Judaici* (with a translation into Spanish), Ciriaca Morano (ed.) *Isaac Newton El Templo De Salomo'n* (Madrid, 1998).

cultures. On this topic, Newton writes: "To the description of the Temple belongs the knowledge of the *Sacred Cubit;* to understanding of which, the knowledge of the Cubits of the different nations will be conducive."[5] Manuel has observed that Newton's scientific reconstruction of the Temple "involves the Sanctuary of God, whose linear measurements, according to Newton's careful computations, doubled from the Tabernacle under the Judges to the Temple under the Kings; and similarly the dimensions of the new Jerusalem under the King of Kings would be double that of royal Jerusalem."[6] Manuel understood this evolving structure as a quantitative expression of supremacy taking possession of the holiest of holies. The meaning of this expansion in terms of world history remains open.

Westfall too writes that the Jewish ceremonial and worship at the Temple was seen as a "type" or figure to aid in the understanding of prophecy. For Newton, Jewish practices and the Book of Revelation are "like twin prophecies of the same things, mutually explicate each other and cannot be satisfactorily understood apart from one another."[7] The Tabernacle of Moses and the Temples of Solomon and Ezekiel were all built to the same plan except that the dimensions of the temples were twice those of the tabernacle, which also lacked some minor structures near the entrance. Newton carefully reproduced Ezekiel's measurements (Ezekiel 40-42; 43:1-7; and 46: 19-24) since he believed that they were the best description of God's intention.[8]

According to Dobbs, the idea of the entire world as the true and real temple of God with the natural philosopher as priest had ancient roots (we can find this idea also in the writings of Philo, Copernicus, Kepler, and Boyle). Newton fitted the details of regional ritual into this concept as well. Thus he argued that the movement of the priestly processions of the Egyptians showed that their theology was based on the science of the stars and when the Jewish priests approached the altar, they circled the fire, lighting seven lamps to represent the planets moving around the sun. Dobbs argues that Newton considered the Jewish religion to be less corrupted than others, thus he consequently maintained that the Jewish structures built to house the sacred flame were a truer cosmic representations than all other Altars or Temples.[9] However, she does not discuss the issue whether the numerical proportions of the Temple, so carefully worked out by Newton in terms of the sacred cubit, entered in any way into his other cosmic calculations, or into his

---

[5] Isaac Newton, *A Dissertation upon the Sacred Cubit of the Jews and the Cubits of the Several Nations; in which, from the Dimensions of the Greatest Egyptian Pyramid, as taken by Mr. John Greaves, theAntient Cubit of Memphis is Determined.* Translated from the Latin of Sir Isaac Newton, reproduced from Birch (ed.), *John Greaves*, (1737), vol. 2.

[6] Manuel, *The Religion of Isaac Newton*, p. 92 and also pp. 162-3. See also: Newton, "Of the [world to come], Day of Judgement and World to Come." ibid, appendix B.

[7] Westfall, *Never at Rest*, p. 346, see also: Yahuda MS. 2.4 f. 46.

[8] Westfall, *Never at Rest* p. 346.

[9] Dobbs, *Janus Faces*, pp. 152-3.

determinations of critical prophetic dates, or whether, conversely, proportions from other studies entered into his determination of Temple structure. She does mention, however, that Newton belonged here to a long tradition, which maintained that God himself gave the "pattern" for the Temple to Solomon via David (I Chronicles 28: 11-19).[10]

Goldish expands on this issue, pointing out that Newton was well informed about contemporary interest in the ancient Jewish Temple. He had a special understanding of the Apocalypse, according to which the *entire* action of the work took place in the Temple of Jerusalem.[11] "The temple," writes Newton, "is the scene of the visions."[12] With the aid of his prophetic lexicon Newton explained "how the images presented are in fact representations of the prophet's experience in the Temple" and how every aspect of the "Temple – its physical layout, vessels and ceremonies – thus became critical to the unraveling of the secrets held in the Apocalypse."[13] Goldish remarks that Newton's historical and interpretive method grew out of material by various Christian Hebraists and in Jewish sources; however, he developed his research in his own style, applying it in new directions. The style can be seen at work in the three distinct levels of interpretation of the prophet's vision. The first level is the concrete scene of the prophet walking in the Temple, the second is the cryptic manner in which this is explained in Revelation, and the third is the meaning of this cryptic description, communicated only to those who understand. Newton's detailed account of Revelation taking place in the Temple is intended to demonstrate a form of Christian worship based on the Jewish ceremonies. The most important connection is Newton's insistence that the form of Church government is based on the synagogue. He even says specifically that most offices of the early Christian church were not instituted by Christ but were "rather in long use in the synagogues"[14] and that both synagogal and Church institutions derive from the Temple.[15]

---

[10] Dobbs, *Janus Faces*, pp. 152-3.

[11] Goldish, *Judaism*, p. 96. Two central manuscripts that give a line-by-line reading of the Apocalypse are MS. Yahuda 9.2, and MS Keynes 5; see also Newton's published *Observations upon the Prophecies of Daniel and the Apocalypse of St. John* (London, 1733).

[12] Newton, *Observations*, p. 451.

[13] Goldish, *Judaism*, p. 96.

[14] Goldish, *Judaism*, 99.

[15] Goldish brings an example of the vision of the prophet in order to exemplify Newton's style of symbolic correspondence between Jewish rites and the Apocalypse and the fact that ancient Jewish ceremonies are the origin of the prophet's vision. Newton interprets and compares the four beasts (lion, ox, a face of a man, and a flying eagle) that appear in Revelation with two corresponding texts cited from Rabbinic exegesis on the ways the tribes were organized around the Tabernacle (notifying that the flags of Reuben stood for man, Judah for lion, Ephraim for ox and Dan for the eagle) (Numbers Ch. 1-2). Finally, the whole scene of the beasts is compared with the vision of Ezekiel of the four faces of the Cherubims. Goldish, *Judaism*, pp. 100-1, The two texts are Revelation 4: 6-7 and MS. Yahuda 1.4 pp. 18r-19r; 154r-155r. I also discuss this issue in the next sections.

Another theme that Goldish takes up is Newton's attitude towards the measurements of the Tabernacle and the two Temples. For Newton, "measuring is a prophetic code for building," more specifically, in Revelation "the prophet is referring to the historical rebuilding of Solomon's Temple by Zerubabbel (Ezra 3), in which various changes from Solomon's Temple were instituted."[16] The constructive aspect of the measurement according to which this three-fold form, the Tabernacle and the two Temples was built, represents also the development of world history. According to Goldish, it is difficult to grasp what precisely Newton meant by this idea of measuring as building.

Did Newton truly believe that the ceremonies and measurements of the Temple described in the prophecies created or predicted future historical events? If so, what might be the mechanism for such an astonishing phenomenon? Goldish could not find anything explicit regarding this subject matter anywhere in the manuscripts. However, he cites Newton on the following point: "The building of a new Temple signifies ye building of ye Church anew or in a new state & by consequence imply her fall from her former condition."[17] He concludes that for Newton the historical/prophetic description of the ceremonies around the Tabernacle represented the establishment of the first Temple, the ceremonies of the first Temple those of the Second and those of the Second shaped the synagogue and the Christian Church, and the corruption of the latter.

In *Prolegomena ad Lexici Prophetici* (MS Babson 434) Newton examines the mathematical relationships between the three structures, trying to reconcile the various sources (Kings I, Ezekiel, Josephus, and the Talmud) in order to present a picture of an absolutely symmetrical Temple. His deep interest in the harmonious spatial relationship between the Tabernacle, Solomon's Temple, Ezekiel's Temple, and the Second Temple is worth looking at. According to Goldish and also Manuel, Newton's purpose in all his endeavors to reconstruct the Temple was to demystify the mystical and magical aspects of the ancient rites. Though he believed that the secrets of the heavens were encoded in the physical form of the Temple, still, in contrast to Hermetic ideas, "Newton's understanding of Temple symbolism is highly concrete and mathematical, spurning other vague allusions to ancient mysteries which Newton abhorred."[18] He concretized the symbolism of the Jewish ceremonies by removing vague mystical ideas from them and situating the Temple in the Prytanaea tradition. For Newton the Tabernacle and both Temples were Prytanaea, known to the ancients and passed down to their descendants under the hieroglyph of temple. The structure of the Temple was a Prytaneum, an exact square, with an exactly square platform, meant to represent the planets in exact symmetry around the sun.[19]

---

[16] Goldish, *Judaism*, p. 102, based upon MS Yahuda 9.2, pp. 25r-28r.

[17] MS Yahuda 9.2 pp. 27r-28r, cited in Goldish, *Judaism*, p.103.

[18] Goldish, *Judaism*, p. 96. On Newton's tendency to reduce vague mystical ideas to matters of fact, see: Manuel, *Isaac Newton: Historian*, p.10.

[19] Goldish, *Judaism*, pp. 94-95, Goldish notes this puzzling symmetry, since it was Newton himself

In Keynes MS. 5, Newton explains in what sense he understood the Temple to be a representation of the solar system:

> In the Apocalypse the world natural is represented by the Temple of Jerusalem & the parts of this world by the analogous parts of the Temple: as heaven by the house of the Temple; the highest heaven by the most holy; the Throne of God in heaven by the Ark; the Sun by the bright flame of the fire of the Altar, or by the face of the Son of Man shin~~e~~<i>ng through ~~it~~ <this flame> like the Sun in his strength; the Moon by the burning coals upon the Altar convex above & flat below like an half Moon; the stars by the Lamps; thunder by the song of the Temple & lightning by the flashing of the fire of the Altar; the earth by the Area of the courts & the sea by the great brazen Laver. And hence the parts of the Temple have the same signification wth the analogous parts of the world.[20]

Castillejo, though saying more than any one else on this issue of the expanding dimensions of the three structures, offers a more mystical and esoteric interpretation. More in line with Dobbs' study of Newton's alchemy, which demonstrates how consistent Newton's thought was in all the areas he investigated, Castillejo finds that a unifying expanding force appears throughout. For Newton, ancient rites hide scientific facts, and vice versa. For example, his way of ordering scientific material in the *Principia* and *Opticks* conceals esoteric notions about the way Providence governs the world.[21] Number symbolism appears everywhere. Newton used mathematical symbolism borrowed from his study of Revelation and the measurements of the Temple, such as the numbers three, seven, eight and ten, to structure his *Principia* (three books, opening with eight definitions, followed by three laws of motion) and more clearly later in the *Opticks* (seven books arranged in three parts opening with eight definitions, eight axioms, and eight propositions).[22] Castillejo also connects this number symbolism to Newton's concept of the historical development of humanity, starting from the two generations emanating from Adam (a sequence of seven corresponding to the corrupt branch of a generation of ten representing Noah's branch), the seven Noachide commandments, the measurements of the Temple, the prophecies, and

---

who established the mathematical basis for the elliptical orbits of the planets, and a whole Jewish tradition, which he chose to ignore, maintained that the Temple structure was not a symmetrical square and thus may have been closer to the representation of an elliptical orbit than the symmetry Newton constructed.

[20] Newton, Keynes MS. 5, Vr. 9. Snobelen was kind to give me his transcription of this manuscript.

[21] Castillejo, *Expanding Force*, p. 41.

[22] Castillejo, *Expanding Force*, pp. 98-104; idem, *A Report on the Yahuda Collection*, p. 8.

alchemy.

According to Castillejo, Newton discerned a triple analogy between the Temple, the Universe, and the politico-mystical events of an ecclesiastical community. In his study of Daniel and Revelation, he interpreted the astronomical, terrestrial, animal and architectural images to signify political and ecclesiastical events. Thus in Revelation, the Temple became the scene of visions, and those "visions relate to the feast of the seventh month: for the feasts of the Jews were typical of things to come."[23] Christ prophesied in terms of the first Temple, which represents the history of the Jewish people, whereas John prophesied in the second Temple, which symbolically represented the history of the Christian community. A careful reading of the sources leads to the conclusion that the second Temple's outer court, though expanded in measurements in relation to the first Temple, was left unbuilt and handed over to the Gentiles or to the Beast, while the saints withdrew within.[24] A similar expansion of structure with a loss of particulars occurs in the relationship between the Tabernacle and the first Temple, as Newton wrote across the outer People's Court in his autograph floor-plan of the Temple of Solomon:

> The same God gave the dimensions of the Tabernacle to Moses & Temple with its Courts to David & Ezekiel & altered not the proportions of the areas, but only doubled them in the Temple, abating the thickness of the walls, which are not reckoned. So then Solomon & Ezekiel agree, & are double to Moses.[25]

As will be further elaborated below, the Jewish Temple or community enjoyed an earlier period when the outer court was complete, followed by a period when it was defective and split; similarly the Christian community had an early period of purity, and a later one of corruption.

From these observations Castillejo concludes that Newton conceived of the Tabernacle, the first Temple and the second Temple as representing a kind of molecular structure that develops in time according to the stage of purity of its worshippers. This molecule functions as a square Temple, which attracts and absorbs more and more external matter into itself, while the human world becomes more and more corrupted through idolatry. The absorption into the molecular structure means that the original structure holds on to its true body from within, as the faithful withdraw inwards; consequently the outer court or shell of the molecule/Temple is abandoned to the impure new arrivals. As the molecule increases its over all volume with each absorption into its structure, it incorporates more and more incoming population, expanding its condensed original symmetrical structure. Thus the molecule grows as it splits

---

[23] Castillejo, *Expanding Force*, p. 33.

[24] Castillejo, *Expanding Force*, p. 46

[25] Newton, Sotheby MS. 263, Babson Collection, also cited by Castillejo, *Expanding Force*, p. 44.

and halves itself internally.[26] In the development of this molecule Castillejo thinks to trace Newton's identification of a hidden historical plan of Providence determining both the re-installation of the true pristine religion through a chosen soul, and the descent of succeeding generations into idolatry. Even more remarkable is Castillejo's suggestion of an analogy between such a process and Newtonian physics.

More specifically, Castillejo compares the original stage of the assumed molecule to Newton's inertial state of a body. The reaction of the pristine molecule to disruption of its inertia by a momentary force (analogous to an historical attack by an enemy) may be described in two ways, either from the inner perspective of the body persisting in its inertia (resisting change in its inertial state due to its *vis insita*) or externally, as if an outer force were attacking the inner structure of the molecule. A reciprocal counter-impulsion ensues as the third law of motion predicts. As we have seen, Castillejo argues that Newton's optics, alchemy and work on prophecy all display similar structures and similar numerological forms (3, 7, 8, 10), and therefore probably belong to a unified system following similar laws and conceived as subject to one and the same expansive force governing the history of humanity and the natural world.[27]

Castillejo's unorthodox study has generally been ignored. The fact that he hardly anywhere explains the steps that conduct his argumentation adds to it. The analogies and number symbolism that he uses are not presented scientifically or academically in the meticulous manner which Newton liked so much. Castillejo's style of exposition resembles rather the Hermetic tradition and the Kaballah (though the number symbolism he uses is different). Newton himself abhorred the mystical aspect of these traditions and saw in them the seeds of idolatry and corruption. The whole of his research was devoted, as Manuel and Goldish have said, to demystifying the mysterious and magical and expounding their scientific basis.[28] Moreover, Newton (like Maimonides) while believing in an esoteric tradition reserved to the wise and moral elite, also felt that the inner secrets of providence should not be written down explicitly in any form lest they be misused by the wicked. This was one of his most severe criticisms of Boyle's publication of alchemical secrets.[29] Whatever is esoteric should remain hidden, and if written out, in a secretive language

---

[26] Castillejo, *Expanding Force*, p. 50.

[27] Castillejo, *Expanding Force*, p. 113.

[28] On the issue of demystifying the mysterious and magical, see the work done on Newton and the Kabbalah. Castillejo, *The Expanding Force,* pp. 65-7, Goldish, *Judaism,* pp. 146-155; idem, "Newton on Kabbalah," *The Books of Nature and Scripture*, pp. 89-103. On Newton's adverse response to Leibniz's kabbalistic tendencies see: Allison Coudert, *Leibniz and the Kabbalah* (Dordrecht, 1995); Alison Coudert, "The *Kabbala denudata*: Converting Jews or Seducing Christians?" R. H. Popkin and G. M. Weiner (eds.), *Jewish Christians and Christian Jews from the Renaissance to the Enlightenment.*, (Dordrecht, 1994).

[29] Isaac Newton, "Newton to Oldenburg, 26 April, 1676," *The Correspondence*, vol. 2, pp. 1-2.

understood by the wise and moral alone, as Daniel wrote it in his prophecy. In turn, interpretation as an exposition of the hidden divine secrets must be presented in scientific terminology, otherwise those who have not reached the level of wisdom and virtue necessary for using it correctly will ignore it, underrate it, or find it unfit, and even worse, abuse the message. Thus Castillejo, though putting his hand on something worth further investigation, and to my mind correct, presents it in what Newton would have considered the wrong terminology.

The approach I would like to suggest would be based on a holistic perspective, taking Castillejo's fantastic ideas seriously, yet attempting to present them in simple analogies relying upon modern scientific ideas, which were of course not available to Newton. I chose such a route to enable the modern reader to appreciate that behind Newton's search for the mysterious lay a strict scientific logic. For him there was no real division between his published scientific work and his mysterious unpublished manuscripts on history, prophecy, alchemy, and the Jewish Temple. As he understood the world, the providential design is present everywhere, yet it becomes visible only to those who have become wise and moral. Such wise men understand the limited ability of their generation to accept and understand the providential truth. For this reason, the wise have kept the meaning of the scientific truth from the wider population, making public only informative religious rites.

Therefore, I assume that Newton believed that the vulgarity of his contemporaries, even of the scientific community, was so great that the wider implications of truth had to be kept secret, else he would not have achieved the credible status necessary to change the ways people thought and acted. Moses spent forty years in the desert working to change the consciousness of his generation. Newton did not have such an extreme situation to deal with; his society was less damaged, and much less willing to hear and accept a spiritual truth. Thus he needed to be very wise, cautious, patient and secretive whilst maneuvering his way to the top. His success is indeed remarkable if we take into consideration all the obstacles that stood in his way. Another reason why he hid all the wider implications of his conclusions, even from his own manuscripts, is rather simple. Newton believed in the words of prophecy, and Daniel had said that only at the end would people be able to understand the work of providence. Newton knew his generation was not ripe for the truth since the end had not yet been reached. Therefore, he, like Daniel, left hidden traces of the wider implications of God's providential truth to later generations to decipher. His time was not the time to be explicit:

> For it was revealed to Daniel that the prophecies concerning the last times should be closed up and sealed untill the time of the end: but then the wise should understand, and knowledge should be increased. Dan 12. 4, 9, 10.[30]

---

[30] Newton, Yahuda MS. 1, p. 1r, see also: Snobelen, "God of gods," pp. 204-08.

II. THE LOGIC BEHIND THE EXPANDING MEASUREMENTS

Newton's work on the ceremony around the Prytanaea and his alchemical studies may help to elucidate his thinking on the expanding proportions of the symmetries of the Tabernacle and Temples. Similar symmetries appear in the rites around the sacrificial fire and are displayed in pure metals. Newton's studies of the original religion were motivated by a desire to uncover the ancient natural philosophy, encoded in the Tabernacle and the Temples. Analogous cyclical structures appear in many of his works, most clearly in his historical and alchemical writings, but also in his cosmogony. The Jewish story exhibited to him the "DNA," so to speak, of the worldly design whilst exemplifying the mathematical lawfulness of its corruption through expansion and loss of structure.

In the discussion that follows I shall try to find a plausible meaning for Newton's notion of a mathematical lawfulness of human corruption in the history of the Israelites as it appears from his work on the Tabernacle and the Temples, returning the analogy of the computer for simplification and adducing supporting textual evidence. As I have argued, according to Newton's system, without God's design and constant sustenance of the world all that existed would be God's spiritual unity, yet no creation can occur in such a unified state. However, God chose to create a certain design and organize material atoms according to a chosen plan, thus making a seemingly separation and plurality between himself and the created world. The insertion of the divine program clearly separated absolute space from the material design, and situated material atoms in separate and distinct places in space according to the chosen wise plan of providence. Before the design was chosen all that existed was one unified wholeness with no divisions; only within creation did plurality, limit, and separation begin to rule a material realm.

What is the design? It is God's choice to express his mathematical and mechanical ingenuity in a material realm (which he also designed according to laws of motion which he specifically chose to suit his absolutes), and his choice to express his goodness and justice by creating man in his own image. Without the divine design the material world would remain in a state of symmetrical nothingness, pure potential for any design. The instructions enacted upon the design cause the unconditional eternal flow of God's energy (called absolute time) to lead atoms continuously through logically designed cyclical patterns. The result is that we perceive separated interacting massive bodies within absolute space. The design (the chosen protoplast) enabled God to create a seeming separation and division in space and in the unified eternal flow due to the nature of mass to resist change of its inertial flow. Without mass nothing could be separated and substantiated. The providential design and the chosen protoplast employ the unconditional eternal divine flow and massive atoms for their own purposes. Without the design the natural tendency of God's eternal flow is to disintegrate all designed structures whilst spreading all material atoms evenly and symmetrically in absolute space. This means that two divine poles are constantly at work; the first is the tendency of the eternal flow to disintegrate and dissolve creation into an equable flow of a unified nothingness, the second is the wise instructions of providence, which create separation, limits, and division in the unified flux whilst using wisely and for its own purposes the nature of mass to resist change of its inertial state, thus making it the carrier and preserver of the design.

Put differently, the design of providence conditions the eternal flow to move through separate cyclical patterns due to God's employment of central spiritual forces, such as gravity and fermentation. These separate cyclical patterns and spiritual forces express God's wisdom, though He could have definitely chosen otherwise. Nothing limits God's free will. The wise plan of creation is a limited expression of his unbounded free will. To my understanding, the design of providence as suggested by Newton employs a remarkable logic. It limits the freedom of atoms to flow equably and symmetrically all around space, obliging them instead to follow the laws and orders of intelligent and informative spiritual forces. The more intelligent and ordered the spiritual order the more restricted and lawful are the movements of atoms in absolute space. In other words, when we observe structure and organization in a designed system this means that its components have lost their total inert freedom. The "willingness," so to speak, of each atom to limit its inert freedom within a structured bodily system is what makes the intelligence of the whole design wiser than its separate parts. Each atom by itself obeys only the logic of the unbounded equable non-structured flow yet when combined according to a designed spiritual program its combination with other atoms makes the whole greater (in terms of intelligence and information) than the sum of its atomic parts.

In Newton's system, God could have chosen to create many designs, each substantiated in various orders, yet the design in effect decided upon is a sign of his wisdom and greatness and skill in mechanics and geometry. The beauty of the providential design is that though it produced plurality and separation in the material world everything obeys very simple spiritual and mathematical laws and the basis of creation is defined by a simple prototype cyclical order that reappears everywhere. In order that the protoplast of creation should remain stable and not dissolve immediately into black putrefaction, the instructions of the design need to be constantly read by the eternal fluxion according to loop instructions (cyclical patterns). Thus the unbounded equable flux constantly rereads and reproduces similar prototypes with massive atoms. The atoms creating the material design are always in flux; this is the mechanism behind God's daily sustenance. God's eternal equable flow keeps the whole material design in constant flux.[31] It is only our limited capability to capture the flux that makes us think that material bodies are fixed. However, although the bodily structures remain stable the atoms reproducing them keep vibrating at different tempos.[32]

---

[31] As the image on a computer screen is never static and the electrons reproducing it keep moving in different directions.

[32] To simplify this rather abstract discussion, take for example the human body. All our cells

Therefore, when divine loop instruction is given, the freedom of fluxion of material atoms is restricted since the atoms must obey the specific instruction. An illustration of this mathematical lawful restriction is the design of a central gravitational force, which forces a flux of atoms to move in circular orbits around it. Once an organized system is being constantly reproduced the freedom of its components is restricted. If some of the instructions are not followed by the components of the system disruption enters the order. Therefore, for any given organized system of atoms, such as the solar system, when an unintentionally impressed force prevents the intended order from running smoothly the disruption will also cause damage to the order of the system. Two possible kinds of damage may occur, though there are other possibilities as well. If the source of the harm is a local external aggressive interaction the system might lose some of its external structure.

However, the more hurtful consequence ensues when many components of the system become less obedient to the instructions of the manual, and the degree of freedom of the components keep growing. Thus the original organization around central forces necessarily expands in space according to a mathematical formula since the central force governing the order has less control over the components, each of which wants to return to the stage of total inert freedom.[33] In other words, when the central organizing force loses its power over the components the intended simple harmonic order disappears and complexity followed by chaos replace the original design.

In the manuscript cited below, Newton develops a similar idea. He says that God's presence is everywhere (what I have called as his equable fluxion spread over space) yet the intensity of his presence differs according to the specific order given at each place (namely, the intelligence of the design of each specific system):

> Tis not ye place but ye state wch makes heaven and happiness. For God is alike in all places. He is substantially omnipresent, & as much present in ye lowest Hell as in ye highest heaven, but ye enjoyment of his blessings may be various according to ye variety of places, & according to this variety he is said to be

---

continue to be reproduced at differing tempos throughout our life. In the course of three years the material constitution of our body is almost completely replaced, yet outwardly we may seem the same. Cells are constantly generated and destroyed though the whole body remains similar to itself. The same cycle happens everywhere, in differing tempos. Today scientists speak about a cycle of heavenly stars even older than our sun. Such changing tempos generated by a divine design upon the equable flux might be what Newton understood to be the heavenly music.

[33] More graphically, the daily divine sustenance of the eternal flow reads the design as a current reads the music of a CD recording. The reading of the design means that the current is no longer free to flow equably in space, rather it accelerates and decelerates according to the program inscribed upon the CD in changing tempos. Yet if the instructions are not faithfully obeyed the components of the system follow their own internal inert freedom. What we hear then if we play the CD is a cacophony.

> more in one place less in another, & where he is most enjoyed & most obeyed, there is heaven & his Tabernacle & Kingdom in ye language of ye Prophets.[34]

This text suggests that Newton differentiated between places in uniform absolute space according to the specific order expressed within the place. The more that place was structured according to the original design of the protoplast of creation (as the Tabernacle) and was completely obedient to God's lawful commands the more such a place could be "enjoyed" by human beings and become for them a "heaven." The providential design is so wise, just and good that it designed human beings in accordance with the worldly design in such a way that if they operate morally as instructed they will experience the world as heaven. Yet there is a strict condition; the manual of the design must be completely "obeyed," otherwise the designed heavenly place will lose its order, gradually becoming complex, and later be experienced as hell and a state of total chaos. Newton explicitly says here that God is also present in hell since his sustenance reaches everywhere. The difference between hell and heaven is that hell is a place that totally disobeys the divine order, and thus hardly any of the nourishment and sustenance of the divine flux enter into it. Hell is dark both in terms of physical light and in terms of spiritual and moral enlightenment since chaos and idolatry are states in which each atom (or human being) has become totally egoistic in its inertness, not willing to share its inert freedom with anyone else.

This, according to Newton's historical narrative, is precisely what happened in world history due to idolatry and it is represented by the expanding measurements of the Tabernacle and first Temple. Thus when we reach the first Temple the concentrated encoded design expressed by the Tabernacle expanded twice, lost some of its structure, whilst outer courts added, [35] as Newton explains in detail in Yahuda MS. 9.2, when interpreting the verses of the Apocalypse:

> *The first Beast was like a Lyon, the second like a Calf, the third had a face like a man & the fourth was like a flying Eagle.* These representations of ye four quarters of Israel are taken from the standards of ye tribes in ye wilderness. For ye frame of ye Temple is analogous to their encamping there. The Priests & Levites encamped next about the Tabernacle answerable to y^e

[34] Newton, Yahuda MS. 9.2 ff. 139.

[35] "The same God gave the dimensions of the Tabernacle to Moses and Temple with its courts to David and Ezekiel and altered not the proportions of the area's, but only doubled them in the Temple, abating the thickness of the walls which are not reckoned. So then Solomon and Ezekiel agree, and are double to Moses." Newton, *Prolegomena ad Lexici Prophetici partem secundam in quibus agitur De Forma Sanctuari Judaici*, note to [fol. 9], in *Isaac Newton El Templo De Salomon*, p. 81

> inward Court of y[e] Temple, & round about them were the twelve Tribes in four quarters to y[e] four winds answerable to y[e] outward Court, & in each quarter were three tribes under one standart as you may read in Numb. 1 & 2. The names of y[e] standarts are not there set down, but y[e] Jewish Doctors by tradition teach {see M[r] Mede} that y[e] eastern standart was a Lyon, the western an ox, the southern a man & y[e] northern an eagle.[36]

The same expanding lawfulness, and loss of structure affected the second Temple, but here we enter a much more detailed and complex situation due to the spread of idolatry among the Israelites, as will be further shown below. The expansion of the Temple dimensions continues also into the future third Temple, even though there it signifies the end of idolatry:

> As the linear dimensions of the Temple under the Kings were double to those of the Tabernacle under the Judges, so those of the City under the King of Kings are double to those of the City under the Kings.[37]

III. The Meaning of the Expanding Measurements and other Prophetic Types

Why does Newton attribute such a great significance to these expanding measurements? What do they tell us regarding the plan of providence? As we saw, Newton associates Ezekiel's measurements of the second Temple and John's of the future Temple of the New Jerusalem with historical events due to occur. When a prophet is told to measure the Temple of God he receives a prediction regarding the constitution of the new Church of God, which will replace the corrupted form of the old one. Newton believed that the prophecies prove that when the Israelites in their idolatry reached a severe form of corruption just before the destruction of their first Temple, only a remnant (144,000) of them maintained enough purity to recognize Jesus as the new messenger of God. Thus they and they only continued to belong to the new form of the Church of God.[38] Newton argues that the Jewish prophets

---

[36] Newton, Yahuda MS. 9.2, 9r, also cited in Goldish, *Judaism*, Appendix F, p.197. Goldish was kind to give me his transcription of this manuscript.

[37] Newton, "Of the [world to come] Day of Judgment and World to come," Yahuda MS. 6 folio 12r-19r, cited in Manuel, *The Religion of Isaac Newton*, Appendix B, p. 135.

[38] See e.g., Newton, *Observations,* "[t]he first temple, illuminated by the lamps of the seven

persistently warned their people of the disastrous consequences of their idolatrous actions. These warnings were also codified in the measurements of the prophecies of Ezekiel, Zachariah, Jeremiah, and later in the Apocalypse. Of the meaning behind these measurements Newton writes in Yahuda MS 9.2:

> Now to this history & constitution of ye Temple (though not understood by ye Jews & Christians of those ages) alludes ye divine Apocalypse. The Temple & Altar & they y^t worship therein, being here opposed to ye outward court, do signify ye Courts of ye Temple & Altar wth their buildings & ye peoples court called ye court of weomen. For by ye Temple y^e [a] {[a] See Josephus Antiq.l.8.p.265,26[6?] & de Bel.Jud.l.6.c.14, p. 916.& l.7.c.10, p.949,950.} Jews frequently understood Zerubbabels Temple alone, y^t is all ye buildings belonging to ye inward Court wherein they had worshipped from Zerubbabels time. The outward Court they called ye mountain of ye house. And hence 'tis called here ye Court without ye Temple. The inward building therefore ye Prophet is commanded to measure, y^t is to build & let ye outward lye wast. *For because builders first sett out the measures of what they build, therefore measuring is used as a type of building.* So Ezekiel to signify y^t ye Temple delineated by him should be built, measures every part thereof wth a reed. So ye Angel for ye like reason measures ye new Jerusalem in ye Apocalyps. And so Zachary sees an Angel wth a measuring line going to measure Jerusalem, & another Angel tells him y^t Jerusalem should be inhabited as towns without walls for ye multitude of men & cattel therein Zech.2. And so Jeremy (chap.21.20) describes building by measuring. *The command therefore to measure ye Temple & Altar & y^m y^t dwell therein (yt is their Courts) & to leave ye outward court unmeasured alludes to Ezekiel's measuring ye Temple in ye Babylonian captivity, & to Zerubbabels rebuilding ye inward part of ye Temple while ye outward court was left out & remained unbuilt & open to ye Gentiles.*[39]

churches, is demolished, and a new temple built for them who will not worship [the beast]; and the outward court of this new temple, or outward form of a church, is given to the Gentiles, who worship the beast and his image: while they who will not worship [the beast], are sealed with the name of God in their foreheads, and retire into the inward court of this new temple. These are the 144000 sealed out of the twelve tribes of Israel and called the two witnesses." Chapter 3, # ix, pp. 467-8.

"We may conceive therefore, that when the first temple was destroyed, and a new one built for them who worship in the inward court, two of the seven candlesticks were placed in this new temple." *Observations*, Chapter 3, # xi, p. 469.

These prophecies also gave additional signs, besides the measurements, regarding the replacement of the corrupted form of the Jewish Church of God with the new Christian Church. The Apocalypse is the most explicit on this replacement. For example when dealing with the numbering and sealing of the servants of God it further supports the message of the expanding measurements:

> For this Temple being not wood & stone but ye living Church of God, ye measuring of it & y^{m} y^{t} worship therein & leaving out ye outward court of unholy Gentiles unmeasured, is a type of ye same kind wth numbring & sealing ye servants of God out of all Israel & leaving ye rest unsealed, & signifying ye very same thing. For as y^{t} signifies ye selecting of a few to be a holy people to God, while ye multitude are rejected: so this signifies ye building up ye elect a spiritual house an holy Priesthood to offer up sacrifices to God by Jesus Christ (as Peter expresses it) while ye outward Court of unholy nations lies spiritually unbuilt. *These therefore are but several types of one & ye same thing, & by consequence this measuring begins at ye same time wth ye numbring & sealing,* y^{t} is presently after ye opening of ye seventh seale; & they who worship in ye measured Temple are ye same wth those saints whose prayers ye Angel in ye solemnity of ye Temple worship presently after ye opening of y^{t} seal offers wth much incense upon ye golden Altar. The sealing of ye servants of God alludes to Ezekiels vision of sealing or marking ye best of ye people in ye commencing Babylonian captivity to be preserved & continue ye remnant of Gods Church in ye times of ye Second Temple. And therefore ye sealed live in ye times of y^{t} Temple, & their sealing as well as ye measuring of y^{t} Temple commences at ye Babylonian captivity. For seeing ye Church never ceases but is built up in a new form & state so soon as demolished in an old one, ye interval between ye fall of ye first Temple & building of ye second is not here to be considered.[40]

Newton's interpretation of the Apocalypse deciphered more types of the Temple. These types and their historical significance are a proof of God's providence. The prophetic scenes of the Apocalypse begin with the Tabernacle and first Temple, go on to the heavenly Temple and end with the prophet prophesizing in the second Temple, whilst the angel measuring out the future Temple of the New Jerusalem. This sequence of events describes the history of the true Church of God from Moses until the Apocalypse, and predict the

---

[39] Newton, Yahuda MS. 9.2, 27r, also cited in Goldish, p. 214, my emphasis.

[40] Newton, Yahuda MS. 9.2, 28r, also cited in Goldish, p. 215, my emphasis.

future of the New Jerusalem:

> The Temple being the scene of the visions I conceive that it remains the same from the beginning to the end, & that in allusion to the times of the Tabernacle or first Temple or those of the second for representing the various states of the Church the things that appear in the Temple are only changed.[41]

In the above text (Keynes MS. 5) Newton describes in detail the major differences between the types of the first Temple and of the second one as they reappear in the Apocalypse in order to delineate the work of providence in replacing the Jewish Church with the Christian Church. The Church of God refers only to those who remain pure and follow the precepts of Noah's religion, whether they are Jews or Christians. Any form of idolatry corrupts God's Church, and consequently God needs to reinvigorate and replace the corrupt form of the Church with a new and purer one. Christ comes as a messenger at a very corrupted stage of the Israelites, as Moses came earlier when they were slaves in Egypt. Newton gives a detailed description of the differing types of the first purer form of the Israelites' Church of God and the later corrupted period when Christ came to reform it:

> And in all cases the Church is represented by the Candlesticks wch appear in the Temple & by the mystical body of Christ in what form so ever he appears & by the four Beasts whenever they appear. In the beginning of the Prophesy untill the seventh Trumpet is ready to sound, & again in the repetition or interpretation of this prophesy from the time that the Temple of God was opened in heaven & the Ark of his testament was seen in it untill the seventh Viall be poured out, the times of the ~~first~~ Tabernacle & first Temple are alluded unto. And in all these visions the Church of God is represented by the golden Candlestick wth seven branches which was placed in the Tabernacle & first Temple. Solomon placed ten such candlesticks in the first Temple, but each of them being sufficient to represent the Church, there is but one wth its seven branches considered in the prophesy. In this Tabernacle or Temple the Church is also represented by the four Beasts & by the Lamb with seven horns whether he appears before the throne & opens the seales of the Book or stands on mount Sion with the 144000 [the pure representatives of the twelve tribes of the Israelites].

> [He goes to describe the second corrupted period]

> In the latter part of the prophesy from the time that the mighty Angels comes down from heaven with the Book open in his hand untill the sounding of the seventh Trumpet, & again in the repetition or interpretation of this part of the prophesy when one of the seven Angels carries John into the wilderness to see the great whore sitting upon her Beast, the times of the Babylonian captivity & second Temple are alluded unto, & the Church of God is represented by the two Candlesticks & <two> olive-trees wch the Jews (out of poverty) placed in this Temple upon their first return from captivity Zech. 4, & by the mystical body of the Son of man in the form of an Angel wth feet as [41r] pillars of fire standing on the earth & sea, & instead of the four Beasts the Gentiles are placed in the outward Court.[42]

The major difference in the types of both periods is the number of candlesticks in the Temple. In the first period there are seven candlesticks; in the later there are only two, with two olive trees called the two witnesses. Another difference is that the outward court of the second Temple is left unbuilt, and the twelve tribes of Israel represented by the four Beasts that stood around the Tabernacle in the wilderness (or stood in the built outward court of the first Temple) are replaced by the Gentiles who contaminate the unbuilt outward court of the second Temple. The manuscript continues to explain the significance of these differences in terms of the hidden work of providence in world history:

> ffor Zerubbabel built only the courts of the Temple & Altar & the <new Court or> weomens court for the use of God's people & left the outward court unbuilt & open to the Gentiles. When the twelve tribes of Israel represented by the four Beasts standing in the outward court, became divided into two parties one of wch is sealed with the seal of God [the 144000] & the other receives the mark of the Beast [all the rest of the Israelites]: the multitude of Israel wch receive the mark of the Beast still continue in the outward court & for their idolatry in worshipping the Beast & his Image as the heathens worshipped their fals Gods & Idols, are called Gentiles.[43]

---

[41] Newton, Keynes MS. 5, 40r.

[42] Newton, Keynes MS. 5, 40r-41r.

[43] Newton, Keynes MS. 5, 41r.

Thus the idolatrous Israelites, who worship God in the unbuilt outward court, can no longer be considered part of the true Church. Newton is adamant on this issue. He says that these idolatrous Israelites:

> are the synagogue of Satan who say they are Jews & are not. They have an outward form of religion & church government & therefore are said to worship in the outward court, but under this form of religion & church government they worship the Beast & his Image & therefore a<re> gentiles. And tho they worship in the outward court of the Temple they tread under foot the holy city & therefore are Babylonian gentiles.[44]

Idolatry creates an outward form of religion which is a corrupted form of God's original religion. God's original religion, in contrast, is constituted by followers who are obedient to the divine commandments:

> When the Tabernacle or first Temple is alluded unto there are 144000 sealed out of all the twelve tribes of Israel <& an Angel filled a Censer with fire of the altar & cast it to the earth>, & this is an allusion to Ezekiels vision in the times of the first Temple ~~when~~ just before the Babylonian captivity when a man was commanded to go through the midst of Jerusalem & set a mark upon the foreheads of the men that signed & cried for all the abominations done in the midst thereof, & to take coals of fire from between the Cherubims & scatter them over the city.[45]

The actions of this pure remnant as well as the idolatrous acts of the corrupt majority influence the future historical development of the Church of God:

> When the ~~second Baylo~~ times of the Babylonian captivity & second temple are alluded unto John is commanded to measure the temple & altar & the court of them that worship therein & this is in allusion to Ezekiels measuring the Temple & altar wth their courts in the beginning of the Babylonian captivity. When the Tabernacle & first Temple is alluded unto the 144000 who are sealed & get the victory over the mark of ye Beast.... & this is in allusion to the dedi~~d~~<c>ation of the Tabernacle & Solomons Temple & signifies a new state of the Church

[44] Newton, Keynes MS. 5, 41r.

[45] Newton, Keynes MS. 5, 41r.

> commencing with the ~~? mark ?~~ sealing of those that get the victory over the mark of the Beast & worship in this temple. And when the second temple is alluded unto the same thing is signified by measuring the Temple & Altar in token of rebuilding them with their courts for those that worship therein: ffor measuring is a type of building Zech. 2.2,4.[46]

This whole scenario is also explicated through the types of the candlesticks:

> And because in the tabernacle & first Temple there was a golden [42r] candlestick with seven branches & in the second Temple two Candlesticks ~~of~~ <& two> olive-trees: the septenary number is applied to the Church when the ~~fir~~ tabernacle or first Temple is alluded unto & the binary when the second.[47]

Here Newton presents a sequential expansion of a divine structure that in many ways recalls the alchemical cycle. During the Tabernacle and the first Temple periods the Church of God was represented by a sevenfold dimension, which had both a spatial and a temporal meaning. In *Observations upon the Apocalypse of St. John* the meaning of these seven dimensions is further elaborated:

> And there were seven lamps of fire burning [in the temple] before the throne, which are seven spirits of God, or angels of the seven churches, represented in the beginning of this prophecy by seven stars.[48]

The seven candlesticks represent the stars, as well as the seven Churches of Asia. However when the Israelites became idolatrous the sevenfold type turned into a twofold one. In the twofold type the whole first Temple edifice of the sevenfold type split into a pure inward court, twice as large as the sevenfold one, and an outward unbuilt court of the second Temple twofold type, which became the place of those who had become corrupted. In Yahuda MS 9.2, Newton describes two additional major types of differences that exhibit the same expansion and loss of structure in the inward and outward court of the first Temple compared to the second. The first concerns the expansion and transformation of the four headed Cherubins of the first Temple

[46] Newton, Keynes MS. 5, 41r.

[47] Newton, Keynes MS. 5, 42r.

[48] Newton, *Observations*, part II, #vi, pp. pp. 452-3.

who were placed at the centre of worship into the four separate Beasts of the Apocalypse located at the four corners of the Temple. The second difference deals with the place of the Ark in the first Temple and in the second:

> These Beasts, Ezekiel afterwards (chap. 10) calls Cherubins & describes them to be full of eyes: so that they resemble the Apocalyptick beasts, but yet wth this difference that one Cherubin with four faces to ye four winds signifies what all four Apocalyptick Beasts (each with but one of those faces) do together, that is, all the four quarters of the Church. Whence they are placed not at a distance from the throne to ye four winds like ye Apocalyptick beasts but in the center even in the throne it self as supporters thereof like ye two Cherubins on either end the Ark.[49]

> In the time of the second Temple when there was no Ark in the most Holy, the book of the Temple was laid up in an Ark or chest in some other place & handed to the High Priest by inferior officers: but how it was taken out of ye most holy in the time of ye Tabernacle & first Temple I do not find recorded.[50]

What is the reason underlying this expansion and loss of structure of the Church of God from a sevenfold to a twofold one? We learn from Keynes MS 5. that this expansion had a hidden historical meaning. The prophecy describes the historical development of the Israelites surrounded by idolatrous nations (called the Dragon) until the destruction of the first Temple and the Babylonian captivity. The actions of the Israelites during this historical period are recorded and represented by a mystical Woman who flees from her enemy (the Dragon). At a certain moment the Woman is influenced by the Dragon and becomes idolatrous and corrupt herself. The Woman then split into two camps, her seed remains pure and becomes the 144000 sealed representatives of the twelve tribes, whilst the rest of the nation, corrupted turn the original Woman into a Babylonian Whore:

> [T]he seven Churches of Asia being seated in the Dragon's kingdom & in the eastern part of the Roman empire, the first Temple with its golden candlesticks may be taken to represent the Church of God within the Dragons kingdom from the beginning to the end of the Prophesy, whether that kingdom be the whole Roman Empire or only the eastern part thereof. And

[49] Newton, Yahuda MS. 9.2, 10r, cited also in Goldish, p. 197.

[50] Newton, Yahuda MS. 9.2, 12r, cited also in Goldish, p. 200.

> by consequence when the Woman flyes from the Dragon & from the temple into the wilderness & ceases to be a part of Christs mystical body, & so is no longer represented by the seven horns of the Lamb nor by the seven candlesticks: the Temple of the tabernacle <with its seven candlesticks> becomes restrained to signify the remnant of the Womans seed wch she leaves behind her in this Temple & in the kingdom of the Dragon who came down to the inhabitants of the earth & sea in the outward court of this Temple.[51]

Newton expounds the role of the Dragon and the splitting of the Woman in the prophecy:

> When the Dragon signifies the whole Roman Empire the Woman signifies the whole Church untill by the persecution of the Dragon~~s~~ she separates in communion from the remnant of her seed who keep the commandments of God & have the testimony of Jesus, & then she signifies a Church in a state of defection schism & apostasy from that remnant, & the remnant signifies the whole true church of God.[52]

In his *Observations*, Newton describes this same historical process together with the measurements of the Temple, but in less detail than the manuscripts cited above. Nonetheless, since here the language is clearer and more precise, I cite the published version as a summation of the historical process of these types:

> And the angel stood, upon the earth and sea, saying, "Rise and measure the temple of God and the altar, and them that worship therein:" that is, their courts with the buildings thereon, vis. The square court of the temple called the separate place, and the square court of the altar called the priests court, and the court of them that worship in the temple called new court: "But the [great] court which is without the temple leave out, and measure it not, for it given to the Gentiles, and the holy city shall they tread under foot forty and two months." This measuring has reference to Ezekiel's measuring the temple of Solomon. There the whole temple, including the outward court, was measured, to signify that it should be rebuilt in the latter days: here the courts of the temple and altar, and they who

---

[51] Newton, Keynes MS. 5., 42r.

[52] Newton, Keynes MS. 5., 41v.

> worship therein, are only measured, to signify the building of the second temple, for those who are sealed out of all the twelve tribes of Israel, and worship in the inward court of sincerity and truth: but John is commanded to leave out the outward court, or outward form of religion and church-government, because it is given to the Babylonian Gentiles. For the glorious woman in heaven, the remnant of whose seed kept the commandments of God, and had the testimony of Jesus, continued the same woman in outward form after her flight into the wilderness, whereby she quitted her former sincerity and piety and became a great whore. She lost her chastity, but kept her outward form and shape.[53]

We need not enter into all the details of the prophecy to realize that Newton is here laying out a divine lawfulness of historical processes similar to his scientific laws of motion and the laws of optics. What is important for my argument is to point out that the reasoning behind these prophetic interpretations is based on the pure or idolatrous actions of human beings. Put differently, these texts set out to show what happens to the original social protoplast of creation when idolatry enters human history. The unadulterated Woman in the Apocalypse symbolizing the Church of God (also the mystical body of Christ) remains in a pure state as long as her people follow the divine commandments. But when the majority of the Israelites commit idolatrous actions this same Woman splits into two camps; the first becomes the seed of the pure remnant of the Israelites chosen for their obedience to the commandments, whilst the rest become the corrupted Whore of Babylon. When this Woman is transformed into the Whore of Babylon she still maintains an outward form of the original religion; however she is no longer connected with the lawful instructions of the pure spiritual seed of the chosen remnant.[54] Similarly, the expansion of the measurements of the inner court and the fact that the outward court remains unbuilt in the second Temple means that the central spiritual force of the nation (represented by the sealed seed of the Woman) has less control over its outer parts, which have become corrupted (the Whore of Babylon). Without God's intervention and his sealing of this spiritual social seed, the Woman would have become totally corrupt and the whole Jewish nation would have disappeared, like other ancient kingdoms when they became corrupt and reached a point of no return, and God did not intervene on their behalf. God's act of sealing the pure remnant of the Israelites was the consequence of his providential wish to reinstall the original religion through a new and wise messenger.

---

[53] Newton, *Observations*, part II, chapter 2, #xvi, pp. 461-2.

[54] In Yahuda MS. 9.2, Newton calls the chosen remnant also the two witnesses, and says that they "allude also to Moses and Aaron ye Prince & High Priest who taught the people ye Law in ye beginning." Newton, Yahuda MS. 9.2, 29r, cited also in Goldish, p. 216.

Idolatry is the action of a human being who disobeys the manual of the divine design. Thus any non-compliant individual action would influence the whole social organization of the Israelites, whether the "outer" majority of the people or the "inner seed" of the nation.[55] According to Newton's historical writings, a similar thing happened in other ancient civilizations, yet these civilizations did not have a covenant with God nor did they have prophecy, so that the cycle of their destruction rotated more rapidly. Indeed as soon as idolatry begins, in whatever form, the capacity of individuals to cause the disintegration of the divine social order increases. The divine social cycle of a society, like all other cycles, is composed first by the stronger tendency of the divine order to impose its structure on the individuals of the society through the Elders, priests and Judges of the society (functioning like a central spiritual force) followed in due course by the stronger tendency within each individual to disconnect from the divine spiritual source whilst becoming materialistic and idolatrous. The result is that the social organization of a kingdom returns to anarchy (putrefaction) unless God decides to intervene.

The measurements of the Temple and Newton's interpretation of the prophesied types show that the central force of the social body represented by the pure remnant is located at the inner court, since the inward court is less influenced by the actions of idolatry and from corrupted external enemies, such as the Dragon. Nonetheless, even the "true spiritual seed" represented by the measurements expands twice, doubling in size from the Tabernacle to the first Temple and from the first Temple to the second Temple. The seed, representing a central spiritual force, is essential for the survival of any living social form. This might also be the reason why Newton insisted that God sealed the seed of the Woman for the future Christian Church, otherwise there might have been a total corruption of the Church of God. Nonetheless, in the divine intervention the sevenfold type of the first period expanded and became a twofold type. The twofold dimension derives from the original sevenfold, becoming the seed of the new reformed Church. The meaning behind the transformation of this expanding seed of the Church of God is not at all clear. Yet, according to Newton, it convincingly describes the historical development of the church of God.

We also know from Newton's interpretation of the Apocalypse that the Temple has two centers, one spiritual and divine hidden from the eyes of the worshippers (the holy place), the other representing the illumination of God's light in the physical universe through the physical burning of the Altar in the middle of the Temple, open to the gaze of all worshippers (representing the sun):

> For Temples were anciently contrived to represent the frame of the Universe as the true Temple of the great God. Heaven is

[55] On the process of idolatry and its consequences to the Israelites, see: Yahuda MS. 21, and Babson MS. 437. This topic is also discussed in Westfall, *Never at Rest*, pp. 354-5.

> represented by the Holy place or main body of the edifice, the highest heaven by the holy place of the most Holy or Adytum, the throne of God by the Ark, the Sun by the bright flame of the fire of the Altar.[56]

Bodies have a similar two-center structure, one hidden and guarded, the other open and interactive with other bodies. The *vis insita* of atoms is never palpably present to anyone yet it is through this passive hidden force that God sustains creation. This force is guarded against invasion. When interrupted it reacts. In contrast, gravity, like the sun, is the illumination of the divine current passing through the hidden *vis insita*, so to speak, to all other massive bodies. The social body of a kingdom is a more sophisticated and complex organization than the atom; nonetheless it exemplifies a similar logic of an inner hidden spiritual force having an outward physical expression. Let us look at this logic in Newton's historical narrative of the systems of governance of ancient kingdoms.

IV. COURTS OF JUSTICE

According to Newton, Noah's religion had a particular social organization where a selected group of Elders executed the divine commandments:

> All nations were originally of the Religion comprehended in the Precepts of the sons of Noah, the chief of w[ch] were to have one God, & not to alienate his worship, nor prophane his name; to abstein from murder, theft, fornication, & all injuries; not to feed on the flesh or drink the blood of a living animal, but to be mercifull even to bruit beasts; & to set up Courts of justice in all cities & societies for putting these laws in execution. In the ancient cities the Judges usually sat in the Gates of the city & were called the Elders of the city & judged of causes both sacred & civil & the father of every family was the elder of the family subordinate to the Elders of the city. This religion descended to Melchisekec & Job & to Abraham Isaac Jacob Moses & the Israelites & to the proselites of the Gate For so the Israelites called the strangers within their Gates who observed the precepts of the sons of Noah.[57]

---

[56] Newton, Keynes MS. 5, # 9, cited in Castillejo, *Expanding Force*, p.33.

[57] Newton, *Irenicum*, Keynes MS. 3, p. 5, quoted also in Goldish, *Judaism*, p. 167.

As long as all the individuals of these societies follow the commandments of God the social organization is loyal to the original divine plan. The social organization of these societies is remarkable. At the beginning none of these societies have kings or civil courts; all the members of the society, including the elders who sit at the gates obey the precepts of God. The elders function as mediators or conduits for the laws of the original religion, as the heavenly bodies are the instruments of the heavenly music. In the social organization, as in the solar system, stability and harmony depend on the nature of interactions between the entities composing the system. As long as the elders and the people of the society follow and internalize the precepts of Noah, the society remains lawful since it is sustained directly through God's lawful commandments designed for humanity. However when idolatry appears the whole social organization is shaken and corruption and expansion enter the society, since idolatrous actions create barriers and distancing among its members, distorting their originally designed loving relationships. Instead of pure loving interactions, as instructed by the second commandment, power, deceit and corruption enter the social body. A similar thing happens in the solar system. As long as the heavenly bodies rotate according to the divine plan there is no loss of motion in the system; however, the moment that external forces disrupt the system loss of motion is inevitable. The manuscript quoted above develops the idea of idolatry as a disruption of the original social/religious design:

> But the Kings of the nations by degrees causing their dead ancestors to be celebrated with sacrifices praises & invocations, the religion of Noah & his sons passed into the worship of dead men & the laws of their courts of justice into the moral Philosophy of the heathens <& the worship of their Gods>>. For Pythagoras one of the oldest Philosophers in Europe, after he had travelled among the eastern nations for the sake of knowledge & conversed with their Priests & Judges & seen their manners, taught his scholars that all men should be friends to all men & even to bruit Beasts & should conciliate the friendship of the Gods by piety, & that a friend was another self, & his disciples were celebrated for loving one another. The religion of Noah & his sons was therefore the moral law of all nations put in execution by their courts of Justice untill they corrupted themselves.[58]

Idolatry turns the divine laws into civil laws that transform the social organization of the society. As corruption spreads the elders are no longer sufficient for the execution of the laws of justice, and a king is needed, followed by the establishment of civil courts. The pure loving relationships

[58] Newton, *Irenicum,* Keynes MS. 3, p. 5, quoted also in Goldish, *Judaism*, p. 167.

amongst the members cannot be preserved, since corruption demands an expansion of the courts of justice. Yet this expansion of the courts of justice, instead of giving the laws more authority over the people, rather point to the weakening of the fabric holding the society together since the laws have now become external to the people. Before corruption entered the society individuals were lawful and loving in an inward manner. This process of outward expansion recalls the splitting of the Woman in the Apocalypse into an outward form of religion, as well as the expanding measurements of the Temples from the sevenfold to a larger twofold type. As long as the social unit remains lawful the law of love is natural among its members and the society follows naturally the sevenfold divine prescriptions. Yet a society is able to endorse such relationships only as long as it remains lawful and worships only God. Idolatry infiltrates the courts of justice of the society, making them less efficient and powerful in executing the moral laws. The court of justice of the elders functioned as a central spiritual force uniting all the people of the society under the law of love. Noah's simple religion consisted of seven commandments, and its courts of justice executed a sevenfold type of laws. After Noah's generation humanity gradually became corrupted. The opposite of lawful love is idolatry. When Noah's sons began to worship dead kings or human beings instead of God they blocked the divine loving current, so to speak, sustained by God on a daily basis, from nourishing the sevenfold structured social unit. An act of restoration of the divine order was called for. When Moses brought his message to the Israelites he needed to install a more powerful court of justice, together with more commandments:

> [After the heathens and the Israelites corrupted themselves] Moses reformed the Israelites from those corruptions & added many new precepts to ye Moral law, writing all down in a book & imposed the whole upon the people of Israel by the covenant of circumcision, & allowed strangers of all nations to live within their Gates without entring into that covenant provided they kept the Precepts of the sons of Noah. And for putting this Law in execution he commanded that the people of Israel should make Judges & Officers in all their Gates. Deut.16.10.
> And these courts continued in Judea till the Babylonian captivity & then were abolished by the Chaldeans (Lament.5.14) & restored by the Commission of Artaxerxes given to Ezra (Ezra 7.25,26 & 10.14) & in the reign of the Greeks were called the Sanhedrim & Synagogues of the Jews. And because the Elders (called Presbyters by the Greeks) judged of things both sacred & civil, they had a place of worship adjoyning to the Court where they sat, & before the Babylonian captivity they had sometimes upon the next high hill an altar for sacrificing & a place for eating the sacrifices

callled the High Place.[59]

Since the days of Moses the court was adjacent to the High Places where the worship around sacrificial fires was performed. The ceremony around the sacrificial fire represented the divine sustenance of creation whereas the courts of justice functioned as the social instruments for executing the divine wisdom. Since God's daily sustenance remains hidden from the majority of people it requires lawful instruments to execute its wisdom amongst the people. Similarly at the Temple God's spiritual and divine sustenance remains hidden from the eyes of the worshippers (in the most holy place), whereas the illumination of God's light is open to the gaze of all worshippers at the central altar representing the sun.

The Church of God is designed according to the ancient religion of Noah and his sons. Ezra's restoration of the Sanhedrin and Synagogues after the return of the Jews from the Babylonian captivity is an indication of its universality. Yet as we learn from Newton's interpretation of the Apocalypse, Ezra's restoration was already a weakening of the sevenfold type of courts of justice of the original religion. We here move into a new phase in the implementation of the divine laws in society within the Church of God. We reach the twofold type, which can only insist on a less demanding lawfulness. Remaining lawful in the twofold dimensional Church of God means:

> When Christ was asked which was the great Commandment of the Law, he answered, Thou shalt love the Lord they God with all they heart & with all thy soul & with all thy mind. This is the first & great commandment, & the second is like unto it, Thou shalt love thy neighbour as thy self. On these two commandments hang all the Law & the Prophets. Mat.22.36. This was the religion of the sons of Noah established by Moses & Christ & is still in force.[60]

Instead of the seven Noachide commandments which Moses reinforced by many additional ones (the 613 commandments),[61] the courts of justice of the Church of God since Christ's first coming execute only the two major commandments, centering mainly on love (the worship of God is even absent from the first commandment in this text). Indeed, according to Newton, Ezra's courts of justice, already a weakened form of the Mosaic courts, directly influenced the social organization of the early Christian Church, which, as we have learned from the Apocalypse, is a reformed two-dimensional Church of God. Thus, the Christian church is a less rigid form of the Jewish institution:

---

[59] Newton, *Irenicum,* Keynes MS. 3, p. 5, quoted also in Goldish, *Judaism*, p. 167-8.

[60] Newton, *Irenicum,* Keynes MS. 3, p. 5, quoted also in Goldish, *Judaism*, p. 168.

[61] On Newton's understanding of the Mosaic law, see: Snobelen, "God of gods," p. 206, note 157.

> Christ & his Apostles therefore instituted no new form of Church government, but continued the ancient form of government which was in use among the Jews in their Synagogues before his coming: And the Christian Churches are nothing else then Synagogues. The Bishop is the Chazan, the Shepherd of the flock, the lamp or candle of the Lord, the Angel or messenger of the Church, the <minister> President of the Synagogue. The Dean seems to be the deputy President of the Council of Presbyters.[62]

There is however an important difference between the Jewish organization and the Christian one:

> Among the Jews I meet with no other courts of judicature then those of the Temple & Synagogues. The Christian Kingdoms have other forms of government & other Courts of judicature for civil causes, & the Chancellours of the Courts of the Christian Synagogues have incroached from the board of Presbyters the judicial power in causes spiritual, so that the Presbyters have long since lost all their authority. The Office of the Deacons is also gone from them being performed by the Overseers of the poor; & the Schools of the Christian Synagogues are ceased & Universities are erected in their stead, as there were also Universities among the Jews in all the cities of the Levites & Priests for instructing youth not in the learning & wisdome of this world but in the knowledge & practice of the law of God.[63]

Could this difference represent the change described in the Apocalypse of the sevenfold number of candlesticks of the first Temple becoming the two candlesticks of the second Temple? Or is it an indication of the corruption of the original institutions of the Christian Church? Let us leave these speculations aside and observe how a similar scientific logic of a spiritual social force losing control over its outward social units as idolatry infiltrates the society reappears in Newton's historical work on the early Gentile kingdoms, before the Church of God split into a spiritual seed of the twofold type.

---

[62] Newton, MS. Bodmer, Ch. 9/10, fol. 4r, quoted also in Goldish, *Judaism*, p. 174.

[63] Newton, MS. Bodmer, Ch. 9/10, fol. 4r, quoted also in Goldish, *Judaism*, p. 175.

V. The Early Gentile Kingdoms

What is remarkable about the cycle of any social order moving towards collapse, according to Newton, is that even its corruption obeys mathematical lawfulness; the tempo and nature of the corruption depend on the character of the idolatrous actions of each individual of the society. Newton's historical narrative of the cycle of birth and death of ancient civilizations obeys this mathematical lawfulness. All these civilizations, originated from Noah and his sons. They all had a similar prototype of worship around the sacrificial fire. The difference between them was in the nature and the speed of corruption of this pristine religion, which was specific to each people. If all civilizations were lawful they would all be similar in their worship of God around the sacrificial fire. Indeed, Newton wrote in a manuscript called "*Theologiae Gentilils Origines Philosophicae*" (The philosophical origins of gentile theology):[64]

> It cannot be believed, however, that religion began with the doctrine of the transmigration of souls and the worship of stars and elements: for there was another religion more ancient than all of these, a religion in which a fire for offering sacrifices burned perpetually in the middle of a sacred place. For the Vestal cult was the most ancient of all.[65]

He adds that when Moses instituted a perpetual flame in the Tabernacle, he was restoring the original worship of Noah and his sons. Noah also learned this worship from his ancestors since this was the true worship instituted by God:

> Now the rationale of this institution was that the God of nature should be worshipped in a temple which imitates nature, in a temple which is, as it were, a reflection of God. Everyone agrees that a Sanctum with a fire in the middle was an emblem of the system of the world.[66]

> [all] mankind lived together in Chaldea under the government of Noah and his sons, until the days of Peleg. So long they were of one language, one society, and one religion. And then they divided the earth, being perhaps disturbed by the rebellion of Nimrod, and forced to leave off building the tower of Babel.

---

[64] For a full description on the manuscript see: Westfall, *Never at Rest*, pp. 351-3, esp. n. 55.

[65] Yahuda MS. 17.3, ff. 8-11.

[66] Yahuda MS. 17.3, ff. 8-11.

> And from thence they spread themselves into the several countries, which fell to their shares; carrying along with them the laws the customs and religion, under which they had till those days been educated and governed by Noah, and his sons and grandsons. And these laws were handed down to Abraham, Melchizedek, and Job, and their contemporaries; and for some time were observed by the judges of the eastern countries.[67]

According to *Chronology*, altars were first erected without temples and it was Solomon who was the first to build a temple later to be followed by other nations.[68] Assuming for the moment that the ancients in fact chose a rite around a sacrificial fire to symbolize God's governance of nature, the choice was ingenious. Very few people have the capacity to grasp the spiritual dominion of God over creation, since God made humans from flesh. Therefore considering the limited spirituality of human beings, the rite around sacrificial fire was an inventive and abstract manner of worship that concisely symbolized God's power over matter. The fire perpetually burning at the center of the sanctum symbolized an eternal divine energy that transforms everything brought into it, yet it itself remains constant. The rite itself signifies the cyclical nature of creation as well as the two polarities within divinity, one hidden transforming every created thing into its inert freedom (in the form of a burning fire) the other observed offering a mathematical structured order like the solar system.[69] The abstract symbol of the perpetual fire at the center of the

---

[67] Following, Newton mentions the idolatry that came out of this religion and then he continues: "This was the morality and religion of the first ages, still called by the Jews, 'The precepts of the sons of Noah.' This was the religion of Moses and the prophets, comprehended in the two great commandments, of 'loving the Lord our God with all our heart and soul and mind, and our neighbour as ourselves.' … and this is the primitive religion of both Jews and Christians, and ought to be the standing religion of all nations, it being for the honour of God, and good of mankind." Newton, *Chronology of Ancient Kingdoms Amended*, Samuel Horsley (ed.), *Isaaci Newtoni Opera Quae Exstant Omnia* (London, 1785). vol. 5, pp. 139-40.

[68] Newton, *Chronology*, pp. 122, 162.

[69] The Hebrew name of this fire is (אש התמיד) e*sh (fire) hatamid (perpetual)*) literally the perpetual fire. Yet the root of *tamid* also gives rise to (התמדה) (*hatmadah*), that is, inertia. I find it perplexing that the name of this divine fire exemplifies the two divine polarities of Newton's system, one pole symbolizing the eternal equable and inertial flow of God's eternal time (*hatmadah*) the other the perpetual fire (*esh hatamid*) open to the gaze of all from which the mathematical harmonic music of the seven planets orbiting around the sun is created. Newton refers to the second functioning of the fire, as follows: "To denote that God is ye author of their life & grow^th^ his appearance is like a jasper & a saphire, that is of a celestial green & red: the green colour referring to ye vegetable rain, water the matter or passive principle out of wch all things grow & are nourished & ye red to ye naturall fire & heat, ye form & life or active principle of all growing things. For ye red is that wch Ezekiel & Daniel in like visions describe by ye colour of amber & appearance of fire, Ezek.1.27 Dan.7.10. Now all this is to represent him that sits upon the throne to be ye eternal author & maker of all things in answer to ye title

world then, is significant and crucial for the running of the worldly design, since, once people start turning idolatrous they worship the elements or a human being instead of God's perpetual order, as it happened to the Greeks:

> [I]t came into fashion among the Greeks... to celebrate the funerals of dead parents with festivals and invocations and sacrifices offered to their ghosts, and to erect magnificent sepulchers in the form of temples, with altars and statues, to persons of renown; and there to honour them publickly with sacrifices and invocations.... They deified their dead in divers manners.[70]

Thus we see that idolatry is more natural to the majority of human beings than the worshipping of God alone, for this reason Moses instituted the first commandment: "Thou shalt have no other gods before me." (Exodus 20:26), and the second:

> The kingdom of the lower Egypt began to worship their kings before the days of Moses; and to this the second commandment is opposed.[71]

Newton declared again and again that idolatry is a lack of capacity of the "common herd" to worship God directly. Therefore, the ancient priests presented to the majority of people a material ritual to reach the spirituality of God's governance:

> As a symbol of the round orb with the solar fire in the center, Numa erected a round temple in honor of Vesta, and ordained a perpetual fire to be kept in the middle of it.... And in the Vestal ceremonies we can recognize the spirit of Egyptians who concealed the mysteries that were above the capacity of the common herd under the veil of religious rites and hieroglyphic symbols.[72]

Further more, God prefers to be worshipped, not for the necessary aspects of his being, such as his omniscience and omnipotence, but for what he

---

& worship wch is afterwards given him." Yahuda MS. 9.2, 7v, also cited in Goldish, *Judaism*, p. 193.

[70] Newton, *Chronology*, p. 121.

[71] Newton, *Chronology*, p. 122.

[72] *Add MS* 3990, f. 1, a passage from the early Book II of the *Principia* taken directly from "Origins of Gentile Theology," also cited in Westfall, *Never at Rest*, p. 434 and previously in this book.

has *done*, that is, for the beauty, goodness, justice and wisdom of his creation:

> Ye wisest of beings requires of us to be celebrated not so much for his essence as for his actions, the creating preserving & governing all things according to his good will & pleasure. The wisdome power goodness & justice wich he always exerts in his actions are his glory wch he stands so much upon , & is so jealous of.[73]

However, it is built into the providential program that all the societies worshipping the true religion (that is, the ancient gentile kingdoms) will go through a cycle that starts with pure worship followed by a process of idolatrous corruption. The providential design has one simple docile way of going right, yet multifarious ways of becoming twisted and complex until it reaches a stage of no return. This troubling time of no return becomes the putrefaction, necessary for a restoration of the same simple religion:

> As often as mankind has swerved from [love of God and man] God has made a reformation. When ye sons of Adam erred & the thoughts of their heart became evil continually God selected Noah to people a new world & when ye posterity of Noah transgressed & began to invoke dead men God selected Abraham & his posterity & when transgressed in Egypt God reformed them by Moses & when they relapsed to idolatry & immortality, God sent prophets to reform them & punished them by the Bablonyan captivity & when they that returned from captivity mixed human inventions wth the law of Moses under the name of traditions & laid the stress of religion not upon the acts of the mind but upon outward acts & ceremonies God sent Christ to reform them & when the nation received him not God called the Gentiles, & now the Gentiles have corrupted themselves we have may expect that God in due time will make a new reformation. And in all the reformations of religion hitherto made the religion in respect of God & our neighbour is one & the same religion… so that this is the oldest religion in the world… All other religions have been set on foot for politique ends, ... [74]

In "Origines" Newton set out to compose a similar story regarding the Gentiles. He adduces evidence that a similar worship around vestal fires had

[73] Yahuda MS. 21, f.2. Cited also in Westfall, *Never at Rest*, p. 355.

[74] MS. Keynes 3, pp. 35-6 quoted also in Goldish, *Judaism*, p. 63-4.

been the oldest in Italy, Greece, Persia, Egypt, and other ancient kingdoms. Yet all these societies very early on deserted this religion and started to worship twelve gods, under various names. Lacking the revelation, which God initiates with Moses and his followers, the Gentiles had only their reason to guide them. For them, the special revealed religion delivered in the time of Moses was not an optional source of additional knowledge about God.[75] Through a study of the origin of the names of the gentile gods and what is told about them Newton concluded that the twelve gods of several people are actually deified ancestors of Noah, his sons, and grandchildren. Though this religion passed from people to people, each used it for its own ends by identifying the gods with its own early kings and heroes; thus each people corrupted the original religion in a slightly different way. Yet all of the peoples' twelve gods had common characteristics. For example all nations worshipped one god whom they considered as the ancestor of the rest. They described him as an old man and associated him with time and the sea. Newton concluded that this old man, sometimes called Saturn and at other times Janus, was the model for Noah, since Janus and "Saturn were but two names of Noah split into two persons."[76] What is interesting in this historical narrative is the assumption that all the gentile people became idolatrous in a similar manner. They misinterpreted the spiritual religious meaning of the abstract ceremony around the sacrificial fire, exchanging it for the worldly scientific truth it symbolized, and began to worship the elements instead of the one God, followed by the worship of twelve of their ancestors, each associated with one of the twelve basic elements:

> That Gentile Theology was Philosophical [i.e. scientific] and referred primarily to the Astronomical and Physical Science of the World System: and that the twelve Gods of the ancient Peoples were the seven Planets with the four elements and the quintessence of the Earth.[77]

From then on each people became corrupted in a slightly different manner. The Israelites, in contrast, were more guarded against this scientific/materialistic culture since Moses was well aware to this problem. In a letter to Burnet from the 1680's, Newton praises Moses for his wisdom:

> Consider therefore whether any one who understood the process of ye creation & designed to accommodate to ye vulgar

[75] On this see: James E. Force, "The Newtonians and Deism," *Essays on the Context, Nature, and Influence of Isaac Newton's Theology*, p. 57.

[76] MS Yahuda 41, p. 2r.

[77] MS. Yahuda 16, p. 1r, quoted in Westfall, "Isaac Newton's *Theologiae Gentilis*," p. 18.

> not an Ideal or poetical but a true description of it as succinctly & theologically as Moses has done, without omitting any thing material wch ye vulgar have a notion of or describing any thing further then the vulgar have a notion of, could mend that description wch Moses has given us.[78]

God the designer intervenes in world history only when the worldly design is about to collapse due to the corruption of the Church of God. Newton believed that the providential plan is not eternal; it is eternal only in the sense that it, like everything within it, is cyclical. As the prophets have prophesied, an apocalypse awaits in the future. Newton's interpretation of the apocalyptic end fits his whole systematic way of understanding the cyclical processes of creation. The end of the cycle is the putrefaction necessary for a better future where humanity will start all over again. Indeed, as Westfall argues, for Newton "the second coming of Christ would bring the final triumph of Christianity rather than a cataclysmic destruction of the physical world." Newton "spoke of it in terms of 'ye establishment of true religion,' and 'the preaching of ye everlasting gospel to every nation & tongue & kindred & people....'"[79]

To conclude: as I have argued, Newton's God created the nature of the material design in accordance with the mathematical qualities of his absolutes. Thus the three laws of motion function as the hardware of God's computer, so to speak, and the mathematical and mechanical order of the solar system worshipped around the Prytanaea and inscribed in the structure of the Tabernacle and the two Temples functions as the material designed order. This divine order can go through three stages: a simple, a more complex, and finally a chaotic. More specifically, the concentrated physical structure of the Tabernacle and the movement of the twelve tribes around it according to the ancient rites of the true religion are God's revelation to Moses and the Israelites of the purest logical and geometrical form of the original protoplast of the material as well as the human design. From creation onward God daily sustains this pure and original order revealed directly only to chosen messenger souls. God's simple providential design involves twelve elements organized in accordance with a most harmonious mathematical order of which the number seven is the most prominent component. Moreover, this geometrical-mechanical design is also inscribed in many differing tempos and sizes all over space, thus assuring that heaven and life on earth will display the wisdom, justice and goodness of the work of God. Not only the Israelites discovered the importance of the number twelve and the number seven. All the gentile civilizations worshipped twelve gods and distinguished the number seven. The number twelve had a scientific significance; it included the seven heavenly bodies, the four elements and the quintessence, the simplest building blocks of

---

[78] Newton, *The Correspondence* (2, n.14) vol. 2, p. 333 ("Newton to Burnet," January, 1680-1).

[79] Westfall, *Never at Rest*, p. 330, citing Newton from Yahuda MS. 1.3, ff. 55-6.

creation. Only later in history, when the Israelites corrupted themselves and were captivated by the Babylonians, the Church of God has organized itself around a twofold lawfulness under the reign of Christ. Nonetheless, the prominence of the number seven was still widespread as we can learn from its reoccurrence in the Apocalypse.

As I understand Newton's world, this protoplast inscribed in the Tabernacle is like a mathematical iterating function, which God inscribes upon the material design each time with a slightly different twist. The bodies of human beings are also designed in accordance with this lawfulness; thus when human beings forget the nature of their simple truth, they stop obeying the manual of the design and corrupt themselves and the whole material order. At the point of no return, just before the collapse of the worldly computer, so to speak, God reveals the concentrated formula of creation and the most basic instructions of the manual to moral and spiritual men, such as Noah, Moses, and Christ. He instructs them to encode the concentrated formula in a religious rite that suits the mentality of their generation. Inscription of the formula in a religious rite centering upon the mathematical lawfulness of the heavens becomes the daily reminder of the manual to the broader, less spiritual population. In Newton's days this notion could be revealed only in scientific works, in difficult mathematical formulas such as the *Principia* and *Opticks*, since the generation was too complex to be able to receive truth in its simplicity.

CHAPTER X

FINALE: WHY DOES GOD HIDE THE HEAVENLY MUSIC

> "Make a joyful noise unto the Lord, all the earth: make a loud noise, and rejoice and sing and praise. Sing unto the Lord with the harp; with the harp, and the voice of psalm. With trumpets and sound of cornet make a joyful noise before the Lord, the King." (Psalm, 98:4-6.)

Newton believed that God wishes us to worship him not for his eternal qualities such as omnipresence or omniscience but for his works:

> And as the wisest of men delight not so much to be commended for their height of birth, strength of body, beauty, strong memory, large fantasy, or other such gifts of nature, as for their wise, good, and great actions, the issues of their will: so the wisest of beings requires of us to be celebrated not so much for his essence as for His actions, and the creating, preserving, and governing of all things according to His good will and pleasure. The wisdom, power, goodness, and justice, which He always exerts in His actions are His glory.[1]

In order that human beings should glorify and worship God correctly they need to learn to appreciate creation for what it truly is. If creation were a matter of necessity, even in the softer Leibnizian determinism of being only a necessity to choose the best, human beings would not be obliged to glorify God for his works since creation would have been but a necessary outcome of his nature. Yet Newton believed that we should admire and love God because he chose under no coercion whatsoever to create a beautiful and wise design that deliberately hides parts of the providential scheme from human beings. The design is partially obscured in order that each person should come to learn individually to appreciate God's love, goodness, and justice through facing directly the moral dilemma with a free will of his/her own. In the Newtonian design, God helps humans indirectly through the law of justice that guarantees that the more a person chooses to do good as God has defined it the more transparent the design becomes for that person. If the providential design were totally transparent to human beings they would not experience free will and a sense of themselves as creators. In order that a will be free it needs to experience the possibility of choosing between alternatives and deciding in favor of one of the several possibilities. Precisely for this reason the

[1] Newton, Yahuda MS. Var. 1 and MS 21, f.2 as cited in Dobbs, *Janus Faces*, p. 87.

providential design is partially hidden, to enable humanity to choose between good and evil. Nonetheless, the design is devised that the more we are able to understand God's hidden providential design the better we are able to appreciate his goodness, justice and love through his works. As a result we understand more fully how to operate the world around us in accordance with the just laws of the design. We also gain a closer and more direct relationship with our Creator since we become like him in our free choices and we become agents similar to God, acting freely without suffering reactions to our spiritual actions.

But how can human beings learn about God's will from his works? If we look at a painting or hear music can we not learn something from the work of art about the skill, talent and free will of the artist? At each moment of the making of a work of art the artist faces many choices and has to decide which to realize and how to do it. The completed work tells us many things about the quality and nature of the artist's choices and free will. For this reason, I believe, it was so necessary for Newton to understand as much as a human being could of the process of choice underlying creation. Newton's humility before God is evident, since nowhere does he pretend to truly know and understand God's motives or rationality, as Leibniz believed humans are able to do. Newton's conclusions about God's providential design are the outcome of an empirical investigation suggesting that a few simple laws govern seemingly very complex phenomena such as the motion of bodies and the work of providence in human history. His empirical conclusions arose from what he considered to be God's direct revelations to humanity. God was the freest being in the world so his design cannot restrict him in any way whatsoever. Therefore his choice to create a world governed by justice, goodness, beauty, and simple mathematical lawfulness is truly a sign of his justice, goodness, beauty and skill in mechanics and geometry, because nothing obliged him to create such an order.

But where does God's free will come from? For us human beings, freedom can only be understood in relation to a restriction or a limitation, or what we call a law (either of nature or a moral law). Creation, in contrast to God's unbounded free will, is already a matter of restrictions and limitations. Therefore, my understanding is that Newton believed that God's absolutes, which condition the material design, give us a clue regarding his free will. The absolutes are God's eternal qualities[2] and as such they are an eternal source of creative energy:

> Time and Place are common affections of all things without which nothing whatsoever can exist. All things are in time as regards duration of existence, and in place as regards amplitude of presence.[3]

---

[2] They exist because God exists. See: McGuire, "Newton on Place, Time, and God," p. 126.

[3] Newton, MS. Add. 3965 Section 13, Folios 545r-546r, #1, translated by McGuire, "Newton on Place, Time, and God," p. 117.

God is one wholeness, one unity, with no divisions or separations. His absolutes are his eternal equable flowing of unlimited (free) energy, the purest potential energy for any design. In this sense the absolutes are the foundational instruments of his free will:

> The most perfect idea of God is that he be one substance, simple, indivisible, live and making live, *necessarily existing everywhere and always,* understanding everything to the utmost, *freely willing good things, by his will effecting all possible things,* and containing all other substances in Him as their underlying principle and place.... *Which always and everywhere can bring to act all possible things, which most freely brings about all things that are best and most accord with reason,* and cannot be induced to act otherwise by error or blind fate.[4]

In order that the unbounded spiritual energy of the absolutes be used to create something separated from himself, God chose (freely under no constraints whatsoever)[5] to design three laws of motion that endowed all created atoms with the qualities of his free will and his eternal absolutes; namely, he made atoms indestructible, mathematical, and equipped with a passive equably flowing force. Assigning the property of mass to all material atoms was a wise move, since the passive inertial force of bodies assures that mass will preserve the created state given to it by the creator. Further more, whenever an external force impels the massive body out of its inertial state the passive force resists and repels in turn, in a mathematical proportion to the intrusive change. The resistance of the inertial mass to a change of its inertial state is an impelled force in itself. Thus whenever any massive body changes its inertial state it automatically becomes an impelling force as regards other bodies. In consequence of this pairing of resistance and impulse the center of gravity of the whole world system always remains in an inertial state in relation to absolute space. It is as if all creation were but a small disruption in equilibrium, floating in an endless, unbounded, eternal, equably flowing spiritual energy.

This balanced state of affairs in the created world would not have obtained if actions and reactions did not cancel each other out due to the nature of mass to resist any change of its original inertial state in proportion to the impelled force. In other words, the original state of all material atoms prior to any designed order is that of total free chaos in which all atoms flow equably along God's absolute space. At this state there are no separations or divisions

---

[4] Newton, MS. Add. 3965 Section 13, Folios 545r-546r, #7, translated by McGuire, "Newton on Place, Time, and God," p. 123, my emphasis.

[5] God is "a being who acts purposively and in accordance with the freely formed dictates of his will." He is in no way a being who is "compelled by necessity and hence incapable of doing otherwise than he does." McGuire, "Newton on Place, Time, and God," p. 127.

in space, it is a state of total nothingness, chaos. Yet this nothingness is not a dead chaos but a very alive one. It is pure potential energy for all creation, or, in Newton terms, the putrefying state necessary for the alchemical creation.[6]

God also chose to create a law of nature according to which bodies obey the third law of action and reaction. This means that if the original inertial freedom of a group of atoms becomes restricted since a certain order is enforced on them by a spiritual hidden force (such as, gravity), there must be a reciprocal effect of this restriction, since otherwise actions and reactions would not be balanced in the worldly design. The balancing effect is what we call the reciprocal effect of gravitational mass. The organization of a group of atoms around a central spiritual force is a sophisticated one in which atoms cancel each other's effect, thus creating one cohesive massive body. In response to the mutual canceling of the opposing forces of atoms around a central force, the specific group of atoms receives a collective stability. By renouncing its total inert freedom and joining a restrictive and limiting ordered system an atom gains greater stability and intelligence than it had in the solitary inert stage, since by giving up its freedom it joined a whole that is more stable than each of its parts: the parts are now joined together around a stable central spiritual force. For any group of bodies this mutual canceling effect can be seen from the center of gravity of that system. In Book I, Corollary IV of the *Principia*, Newton writes:

> The common center of gravity of two or more bodies does not alter its state of motion or rest by the actions of the bodies among themselves; and therefore the common center of all bodies acting upon each other (excluding outward actions and impediments) is either at rest, or moves uniformly in a right line.[7]

There is always a balance between all actions and reactions in the world in relation to God's absolutes that is, in relation to God's equably flowing spiritual energy. This balancing effect is a sign of God's justice and also a wise decision and employment of his unconditional equable flow. The equable flow is the highest stage of equilibrium possible that can be organized; any sort of ordered design interferes with this inert equilibrium. God's chosen laws of motion automatically balance actions and reactions around differing centers of gravity. Each organized system we observe in the world is organized around a differing central spiritual force, which employs this mutual canceling of balancing effects to its own ends, achieving a temporary equilibrium similar to the eternal flux. Yet the equilibrium of any created system is always

---

[6] Analogous to an eternal reservoir of unconditional energy, as the vacuum is described in quantum theory.

[7] Newton, *Principia*, p. 23.

temporary since it is a material limitation of a boundless divine spiritual energy and in due time will return to its spiritual source and to a more stable stage of an equably flowing equilibrium. Created material systems differ from each other in the length of their cycle and in their frequency in relation to the eternal flux. This is the meaning of gravitational mass. Bodies are organized systems of atoms differing in their gravitational mass, each creating a different time cycle in accordance with the organization of its mass around a central spiritual force.

If nothing but God's equable fluxion existed there would be no meaning to cyclical time. People could not construct clocks or watches if there were no gravity, because God's equable duration never exhibits different beats and its rhythm is always that of a unified peaceful flow with no vacillations. It is difficult for human beings to grasp this abstract peaceful energy called divine duration, because it cannot be quantified or measured. The only way human being's can learn about God's eternal abstract time in which nothing happens is through the concept of the center of gravity of the worldly system. This center, Newton tells us, moves in the same constant flux as God's duration and in a straight line in relation to absolute space, so that the velocity of the equable flux of the center of gravity of the whole world can be measured in relation to God's absolute space. Yet God's eternal mathematical time is too abstract and spiritual for most people to comprehend from the concept of an inertial mass. Fortunately, creation is less abstract than God's duration and we can learn from his works about his abstract time through the concept of gravitational mass.

Without gravitation human beings can not measure time. It is thanks to gravitational forces that many cyclical times are created, which human beings, in turn, can measure. To give a few examples: the diurnal orbiting of the earth around its axis, the annual duration around the sun, or the life cycle of vibrating atoms. The second part of Book I of the *Principia* deals with many aspects of measuring this cyclical time in relation to gravitational forces; for example in Proposition XXXVII, Problem XXVI, Newton sets out to "define the times of ascent or descent of a body projected upwards or downwards from a given space."[8] Pendulums also exhibit this cyclical aspect of time, which depends upon gravity. Thus we see that the two aspects of mass, inertia and gravity, are connected through the two differing aspects of time – the equable flowing eternal duration and the cyclical time of the cosmos. The first material aspect, inertia, is more abstract and difficult to capture. It is a material passive state of total freedom with no restrictions or limitations and is associated with the eternal and equable flow of absolute time. It is the first order of lawfulness, which God chose for his design. Gravitational mass, on the other hand, is a more limiting aspect of matter since it is already a second mathematical lawfulness imposed upon matter. In contrast to inertial mass, gravitational mass is no longer associated with God's eternal duration but is related to a created cyclical time. In a more fanciful language we can say that the gravitational lawfulness along with the specific protoplast God chose to create

---

[8] Newton, *Principia*, p. 100.

is the divine music of the spheres. In God's creation each organized material system plays a different cyclical music according to the instructions of its central spiritual force, as reflected in the beliefs of alchemy.

In the alchemical world cycle a spiritual seed (such as a gravitational force or fermentation with a specific tempo) is implanted at the stage of material putrefaction (i.e. a stage of material inertness). The spiritual seed organizes the putrefied matter according to specific divine instructions, creating pure metals such as gold. A series of corruptive stages follows until the metal again reaches a stage of putrefaction, to start a new cycle. Many cyclical systems with differing frequencies and a differing cyclical time are constantly imposed on chaotic stages of matter. According to Newton, the worldly design itself has such a cyclical rhythm although human beings cannot know its length unless the Creator reveals it to them. The reason behind this cyclical phenomenon is that if the world were eternal it would have to be in a stage of total equilibrium (the only stage where there are no repetitions of an order) but then no creation or design could be realized. Clarke, writing on behalf of Newton, implies the cyclical nature of the design:

> The present frame of the solar system, for instance, according to the present laws of motion, will in time fall into confusion and perhaps, after that, will be amended or put into a new form. But this amendment is only relative with regard to our conceptions. In reality with regard to God, the present frame, and the consequent disorder, and the following renovation are all equally parts of the design framed in God's original perfect idea. 'Tis in the frame of the world, as in the frame of man's body; the wisdom of God does not consist in making the present frame of either of them eternal but to last so long as he thought fit.[9]

To illustrate these rather abstract ideas we may think of the created world as a large orchestra playing the music of the spheres. Newton himself uses this analogy in his classical scholia, which are his additional commentaries to the *Principia* (never published) written in the 1690's.[10] In each scholium Newton tracks the classical source of his physical findings, especially that of gravity. The additional scholium for Proposition VIII relates directly to the theme of the music of the spheres. He argues there that Pythagoras discovered the inverse-square relationship of harmoniously vibrating strings and extrapolated from it the ratio of the celestial harmony:

---

[9] Clarke's second reply #8, *G.W. Leibniz: Philosophical Papers and Letters*, p. 681.

[10] In these scholia he points out that the ancients have anticipated his physical discoveries. Though he did not publish these historical commentaries they show that Newton gave preference to the music of the spheres and to the Pythagorean tradition.

> By what proportion gravity decreases by receding from the Planets the ancients have not sufficiently explained. Yet they appear to have adumbrated it by the harmony of celestial spheres, designating the Sun and the remaining six planets, Mercury, Venus, Earth, Mars, Jupiter, Saturn, by means of Apollo with the Lyre of seven strings, and measuring the intervals of the spheres by the intervals of tones. Thus they alleged that seven tones are brought into being, which they called the harmony diapason, and that Saturn moved by the Dorian phthong [voice or mode], that is, the heavy one, and the rest of the planets by sharper ones (as Pliny, bk.1, ch.22 relates, by the mind of Pythagoras) and that the Sun strikes the strings. Hence Macrobius, bk.1 ch.19 says: "Apollo's Lyre of seven strings provides understanding of the motions of all the celestial spheres over which nature has set the Sun as moderator." And Proclus on Plato's Timaeus, bk.3, page 200, "The number seven they have dedicated to Apollo as to him who embraces all symphonies whatsoever".... But by this symbol they indicated that the Sun by his own force acts upon the planets in that harmonic ratio of distances by which the force of tension acts upon strings of different lengths that is reciprocally in the duplicate ratio of the distances. For the force by which the same tension acts on the same string of different lengths is reciprocally as the square of the length of the string.[11]

Let me expand upon this analogy and suppose that the providential material design is analogous to the playing of an orchestral fugue, where the subject is taken up and played on different instruments modulating into different scales. Several basic harmonies characterize each scale – the third, the fourth, the fifth, etc. Each of these harmonies can be mathematically described in simple integral ratio, and thus might be imagined to correspond to the basic heavenly design (the protoplast). The composer uses the scales to form different musical orders. In the worldly design the scales are like an ingenious mathematical and musical order that God, the composer, has inserted in various rhythms upon matter. The concert goes on being played since God sustains creation constantly with his eternal equable flux. He is the conductor and the musician who channels his eternal flow through secondary causes such as gravity. For this task he has designed many central forces that bestow different tempos and frequencies upon the equable flow. All God needs to do after creation is to emanate his eternal flow, which he does anyway, and the result is a divine, harmonious music. It is as if God created out of his eternal

[11] Newton, "Classical Scholia," written for Proposition VIII, cited previously in the book and also in: James, *The Music of the Spheres*, p. 164.

equable flux a number of gentle musical instruments each playing its theme in harmony with the rest.

What is so admirable about God's design, according to my understanding of Newton's system, is that it was specifically chosen in order that human beings, who are also part of the design, would be able to enjoy God's goodness, justice, musical harmonies, and mechanical skill if and only if they completely obey the manual of the design. However, they will be able to understand the laws only if they make mistakes and mar the divine instruments; then they are given another chance to start again, having the historical recollection of their predecessors' mistakes. The tuning of the instruments in an orchestra could be a metaphor for the correcting of the mistakes arising from wrong, i.e. idolatrous, individual actions. But God's amending of the worldly design has nothing to do with this kind of reparatory action that may be performed by a human being. The decision of providence to intervene is a free choice; God also listens to the music. Being the partially hidden conductor and composer he has the right to correct the instruments and the musicians periodically, when the cacophony becomes intolerable. This recurrence of human error that almost totally ruins creation, and the putting right by God is always followed by a more explicit divine revelation of the secrets of the providential design and the laws of creation.

Think of a great orchestra, each instrument designed to be played in a specific way. A part is given for each instrument. If God wished he could play all of the music himself. But for a reason we do not know God has decided otherwise, and has chosen to create musicians, similar to himself (for he created man in his own image) to play the divine instruments, while he partly hides his role as the composer and conductor and the maker of the harmonic laws. God creates musicians knowing in advance that they will choose to play another music than the one intended and designed specifically for each one. To illustrate the freedom God gave to human beings and the necessity they are under to follow the manual if they want to enjoy the divine music, let us look at the following scenario. A conductor is necessary so that the harmonies and rhythms produced by each musician in the orchestra will fit together. In our story the conductor has half concealed himself and has given the players much freedom that they are not even conscious of their role in the concert. Let us suppose that one of them suddenly feels disinclined to follow the notes assigned to him by the composer, or that he disagrees with the conductor's concept, or that he wishes to play leader of the first violins instead of sitting far back among the seconds. The order and harmony of the performance would be destroyed. (Our musicians are both players and audience). If the free spirited violinist influences other players, cacophony and anarchy will ensue and the conductor will be relegated to near invisibility.

Idolatry is such an onset of anarchy. God has created a sophisticated and harmonious design for human beings to live in, yet he purposefully gave them the freedom to disobey the partially hidden conductor who is also the composer and the craftsman of the music, the instruments and the musicians. What is rather startling in Newton's historical narrative is that the chosen wise messengers who reveal the providential secrets at crucial moments in history are always instructed to hide the true scientific meaning of the design from the majority of the people, who are still corrupt. Only the wise and moral are able to know the depth of the truth. Thus Pythagoras hid his findings from the vulgar since

> [t]he Philosophers loved so to mitigate their mystical discourses that in the presence of the vulgar they foolishly propounded vulgar matters for the sake of ridicule, and hid the truth beneath discourses of this kind.[12]

Moses also hid the truth from the vulgar, but for different reasons. Understanding the spiritual limitedness of the majority of people he decided to present the truth in a form that was not beyond their moral capacity. God gave human beings free will, knowing that they would most likely be tempted to chose to disobey the commandments. Thus there is wisdom in hiding the truth from those who do not understand. Nonetheless, if people were not granted free will they would be, in Newton's system, like passive matter, totally obedient to the divine laws and unable to fully appreciate creation for what it is or the goodness and wisdom of the creator. The two differing systems of laws, the natural and the moral orders, differ in this sense. Matter is passive in its obedience to the balancing of causes and effects whereas human beings have freedom to chose which route to take. It is human beings' disobedience to the providential order, which makes the simplicity of the divine plan become complex, even chaotic, enabling cyclical time to enter the ordered design. Without the anarchic and idolatrous tendencies of human beings nothing would bring creation from its golden state into a more complex iron one followed by a destructive chaotic period.

Matter by itself has no say about the order given to it. It is totally passive. With human beings the situation is different. Their existence is encompassed by free will, and for this reason they will be judged for their actions at the Day of Judgment. Yet a potential freedom is also contained in the material laws of the design. As I have argued, the laws of motion, including gravity, leave great freedom for matter to be organized according to whichever initial conditions are chosen. The freedom of matter does not end there. God chose a most remarkable harmonic order, but within its laws he left infinite possibilities for its simplicity to become complex once its temporary equilibrium is destroyed by any idolatrous action of a human being or a material interaction that causes loss of motion. Each individual in the world can influence the possible complexity into which the simple design may develop, as we can infer from Newton's historical narrative of the corruption of the ancient kingdoms, that of the Israelites, and that of the Christian kingdom.

---

[12] Newton, "Classical Scholia," written for Proposition VIII, cited also in: James, *The Music of the Spheres*, p. 165.

God made human beings in his own image, so he presumably also gave them his creative powers, together with his free will to choose which material order to create and substantiate. This is God's goodness and his greatest gift. But it is only through errors and corrections that human beings can learn the providential order and come to love the true conductor and composer of the worldly design. As human beings become aware of the results of their mistakes and understand how to put them right, they appreciate more profoundly the wisdom and goodness of the Creator.

Put differently, in Newton's historical narrative, as long as Adam and Eve were totally obedient to the manual they could never come to know the providential design. They completely enjoyed heaven in ignorance, since they had not eaten from the tree of knowledge of good and evil. They did experience freedom since they internalized the divine laws, yet they did not experience a freedom of choice. Their will was free yet not conscious of its freedom. Only after Eve ate from the apple which gave her the knowledge of good and evil did humanity start experiencing freedom of choice; yet by choosing wrongly people lost their freedom as they made heaven into hell. From then on, in Newton's narrative, the generations fell one after another until Noah was chosen by God (because of his moral character and ability to prefer the good) to reveal some secrets regarding what is behind the partially hidden providential plan. However, he was instructed to hide the true scientific meaning from other people and disclose it only in a religious rite around a sacrificial fire. The generations that followed again fell, and the Gentile society (especially Egypt) rapidly became corrupted, because the scientific knowledge of the twelve elements and seven planets that their priests had received from Noah tempted them to misuse their spiritual powers against their own people.

God then chose Abraham – still according to Newton - because of his search for one God and his abhorrence of idolatry. Abraham received more secrets than Noah had been granted and was instructed to transmit them only to his direct offspring and not to any of the people around him. When we come to Moses we reach a peak in the scientific revelations of creation, yet Moses was clearly instructed to hide the deep meaning from the vulgar majority, who preferred to worship a golden calf rather than to wait for him to bring them their true spiritual and scientific inheritance. Nonetheless, God and Moses, caring for their people, decided to give them another chance, protecting them from idolatry by giving them many more commandments to help them to remain spiritual. Further, Moses and afterwards the prophets kept warning the Israelites by means of visions that foretold the likely outcome of their good versus their idolatrous actions. Yet the prophetic warnings were always given in a coded language known and understood only by the wise, since this knowledge needs to be kept in secret to avoid vulgar abuse.

The providential truth, in Newton's narrative, cannot become explicit to those who have not reached a spiritual and moral stage where they are capable of using it carefully and maturely. One might suppose that the Israelites would have remained spiritually defended by all this reinforcement, yet that was not the case. Although the Israelites were given so much encouragement to become the guardians of the true religion they nevertheless fell. God's mercy once again went to work and Christ came to restore the truth. By then, according to Newton, only a few of the Israelites remained pure enough to recognize the truth he delivered. Christ also took the people's suffering upon himself, knowing that his many merciful actions would partially atone for the sins of those who followed him. With the direct intervention of God he re-established the true religion given to Noah, discarding the many Mosaic commandments and preserving only the two central ones. But what actually happened after his departure was the opposite of his intentions. In succeeding ages the Church placed him on a level with God and the Holy Spirit, thus instituting an idolatrous religion, which Newton so greatly abhorred. The fall of the Christian generations continued until the providential truth again became more explicit and open to the broader population in Newton's time. Newton was chosen to pursue Kepler's work on the heavenly heliocentric design in an explicit mathematical form in the *Principia* and to reveal the hidden design of light in the *Opticks*. Yet the time was not yet ripe to reveal the whole providential order to the world; therefore, Newton remained silent on many issues.

Within the divine scheme all human beings are made of flesh. They are bounded by both sets of laws, the material and the moral/spiritual. They are endowed with a bodily design, which they can either operate in a simple, intelligent and divine way remaining free from material bondages or mishandle and corrupt it through idolatry. God's truth is simple; however, as I have argued throughout this book, according to Newton it is spiritual, mathematical, and very abstract so that it can be fully understood only by a spiritual, wise, and moral person. The mind and heart of an individual have to be capable of being tuned into the divine frequencies, else he will not enjoy the music of the spheres. The design is arranged in such a just way that for every immoral action there is a physical toll. An immoral person is one who has corrupted his sensorium to such a degree that he will never be able to capture and observe the reappearance of the simple harmonic order hidden within every material system. His sensorium has become caught up in the material maze and can no longer be tuned into the gentle harmonic and mathematical spiritual frequencies of the divine music organizing all material systems.[13]

God has equipped all human beings with a pure sensorium like his own. He has designed the world order so justly that humans can capture the divine frequencies directly only when their own sensorium is pure. The divine frequencies are not sensed once the sensorium becomes unfit and too coarse to be tuned into these frequencies. According to Newton's system, any person who commits idolatrous actions, even simple ones such as connecting directly with a sensible object will coarsen his sensorium in due time. For such a person

[13] Modern science illustrates this idea. A human being cannot see infra-red rays with the naked eye but when he puts on specially designed infra-red spectacles which transform rays into visible light he can perceive them. The infra-red frequencies are there all the time but without the right equipment we are blind to their presence.

reality becomes material and vulgar and his desires become materialistic. His sensorium is directly influenced by matter and the more he becomes idolatrous the more complex will be his internal and external world, since he will be caught up in the material maze and will not be able to comprehend the spiritual and mathematical simplicity governing it. Such a person will not be able to accept the simple and harmonious spiritual and mathematical design proposed by Newton. Thinking "rationally" this person will conclude that someone like Newton who has discovered so many connections between the material and spiritual orders is probably hallucinating. Moreover, he will not feel obliged to choose the divine spiritual goodness since it is no longer his internal reality. His materialistic tendencies will probably influence him to perform more anarchic and egoistic actions since the good for him is not the good of the ordered design. The wise, in Newton's sense, in contrast to the vulgar, is one who has become as skillful and virtuous as God. Therefore, the wise, like Christ, will even correct the mistakes imposed upon the design by acting against the third law of motion (the reciprocity of actions and reactions, functioning in the moral realm as a law of justice) choosing to be merciful when retribution is apparently called for.

* * *

Giucciardini has observed that Newton's efforts to link his mathematical methods with those of the ancients are opposed to Leibniz's desire to stress the novelty of his work, and that in general, throughout their research, Newton's dominant themes were continuity and restoration, Leibniz's innovation and revolution.[14] This observation has a puzzling aspect. When we compare the two men's endeavors in their time, we can not help observing Newton's great activity and achievement as opposed to Leibniz's unconcern. Newton's spiritual beliefs are no less puzzling or easy to grasp than Leibniz's and on many points the two share intuitions. Both were committed to their ideas and felt they must convey them to their contemporaries. The difference in communication, however, is enormous. Newton published only what the public could receive, the rest remained concealed and hidden. As Snobelen has observed, Newton "knew the great damage the stain of heresy would do to the cause of his reformation in natural philosophy." All along "he knew a time would come when this would not be so" but until then "he was too much a man of the world not to realize that the day has not come."[15] Leibniz, in contrast, believed that the moment rationality was expressed in the world, the world itself was transformed for the better.

---

[14] See: Niccolo Guicciardini, *Reading the Principia*; Snobelen, "Mathematicians, Historians and Newton's *Principia*," *Annals of Science*, vol. 58 (2001), pp. 75-84.

[15] Snobelen, "Isaac Newton: Heretic," p. 419.

The difference is enormous in another important aspect, connected to their ways of communicating. Newton's understanding of the resistance of the body's *vis insita* to an acceleration is opposed to Leibniz's understanding of the supreme governance of spiritual lawfulness upon material perceptions. In most activities Newton engaged, he maneuvered to face the least resistance from his peers and opponents, since he perceived the defensive mechanisms of human beings as obeying similar laws as material systems that resist change to their inertial state. Any mental defense of the traditional way of thinking, in such a context, will necessarily be an attack on the innovator. This, I believe, was the motive behind his silence, he brought to the public only what it could enhance.[16] Leibniz, on the other hand, felt obliged to share his innovative discoveries with his contemporaries without taking into consideration their spiritual capability and probable hostile reaction to innovative ideas, since in the Leibnizian system a rational distinct thought overpowers and transforms obscure perceptions.

[16] There is a Hebrew saying that sums up this idea: סייג לחכמה שתיקה (silence is a fence around wisdom). Newton's reluctance to publish his innovative ideas in mathematics show his awareness of the human tendency to reject and resist something new unless the innovator is so socially distinguished that hardly anyone dears to challenge him or call his innovation non-scientific. An analogous structure can be found in the laws of motion in the Principia. When a body's *vis insita* is interrupted the body resists the change and reacts accordingly. Thus in any interaction between two bodies the least noticeable change will occur within the massive body, whereas the smaller body will be forced out of its inertial path without delay. Newton experienced this simple truth during the 1670's, when he was constantly interrupted by the many objections of his colleagues to his optical findings. An example of his sensitivity to being challenged is his letter to Oldenburg regarding the critical response of a Mr. Linus to his experiments in optics. Newton writes: "I see I have made my self a slave to philosophy, but if I get free of Mr. Linus's business I will resolutely bid a dew to it eternally, excepting what I do for my private satisfaction or leave to come out after me. *For I see a man must either resolve to put out nothing new or to become a slave to defend it.*" "Newton to Oldenburg," (18/11/1676), The Correspondence, vol. 2, p. 183, my emphasis.

BIBLIOGRAPHY

NEWTON'S WORKS IN MANUSCRIPT

Babson MS: Babson College, Mass.
Dibner Collection: Smithsonian Institution Library.
Keynes MS: Cambridge University: King's College.
ULC: University Library, Cambridge.
Yahuda Manuscripts: Jewish National and University Library, Jerusalem.

NEWTON'S WORKS IN PRINT

"The New Theory about Light and Colors," *Philosophical Transactions of the Royal Society*, no. 80, (1672), pp. 3075-87.

Chronology of Ancient Kingdoms Amended (London, 1728).

Observations upon the Prophecies of Daniel and the Apocalypse of St. John (London, 1733)

A Dissertation upon the Sacred Cubit of the Jews and the Cubits of the Several Nations.... Thomas Birch (ed.), *Miscellaneous Works of John Greaves* (London, 1737), vol. 2, pp. 405-33.

Horsley, Samuel (ed.), *Isaaci Newtoni Opera Quae Exstant Omnia* (5 vol's; London, 1779-1785).

Prolegomena ad Lexici Prophetici Partem Secundum in quibis Agitur De Forma Santuarii Judaici, Ciriaca Morano, (ed.) *Isaac Newton El Templo De Salomo'n*, (Madrid: Consejo Superior De Investigaciones Cientificas, 1998).

The Principia: translated by Andrew Motte (New York: Prometheus Books (Great Mind Series), 1995).

Cohen, I.B. and Whitman, Ann. *Isaac Newton, The Principia: Mathematical Principles of Natural Philosophy, a New Translation by I. Bernard Cohen and Anne Whitman, Assisted by Julia Budenz.* (Berkeley: University of California Press, 1999).

Opticks (New York: Dover Publications, 1952).

Cohen, I. B. and Schofield, R.E. (eds.), *Isaac Newton's Papers and Letters on Natural Philosophy and Related Documents*, (Cambridge: Harvard University Press, 1958).

Hall, A.R. and Hall, M.B. (eds.), *Unpublished Scientific Papers of Isaac Newton* (Cambridge: Cambridge University Press, 1962).

Turnbull, W. H., J.F. Scott, A. Rupert Hall and Laura Tilling (eds.), *The Correspondence of Isaac Newton* (7 vol's; Cambridge: Cambridge University Press, 1959-77).
Whiteside, D.T. (ed.), *The Mathematical Papers of Isaac Newton* (8 vol's; Cambridge: Cambridge University Press, 1967).

LEIBNIZ'S WORKS IN PRINT

Ariew, Roger and Garber, Daniel (eds.), *G.W. Leibniz: Philosophical Essays* (Indianapolis and Cambridge: Hackett, 1989).
Child, J.M. (ed.), *The Early Mathematical Manuscripts of Leibniz* (Chicago: Open Court Publishing Company, 1920).
Dutens, L. (ed.), *G.W. Leibniz: Opera Omnia* (Geneva, 1768, reprinted Hildesheim).
Garber, Daniel and Sleigh, Robert C. (eds.), *De Summa Rerum: Metaphysical Papers (The Yale Leibniz)* (New Haven: Yale University Press, 1992).
Gerhardt, C.I. (ed.), *G.W Leibniz: Mathematische Schriften* (7 vol's; Hildesheim: Olms, 1854 reprinted 1975).
Gerhardt (ed.), *Die Philosophischen Schriften von G.W. Leibniz* (7 vol's; Hildesheim:Olms, 1875 reprinted 1978).
Guhrauer, G.E. (ed.), *G.W. Leibniz, Deutsche Schriften* (2 vol's; Hildesheim: Olms, 1838-40 reprinted 1966).
Leibniz, G.W. *Gottfried Wilhelm Leibniz, Sämtliche Schriften und Briefe* (Dramstadt und Berlin: Deutche Akademie der Wissenschaften zu Berlin, 1923).
Klopp, Onno (ed.), *Die Werke von Leibniz* (11 vol's; Hannover: Klindworths, 1864-88, volumes 7-11 Hildesheim: Olms, reprinted 1970-73).
Austin Farrer (ed.), *Theodicy: On the Goodness of God the Freedom of Man and the Origin of Evil*, (Illinois: Open Court, 1990).
P. Remnant and J. Bennett (eds.) *New Essays on Human Understanding* (Cambridge: Cambridge University Press, 1989).
Loemker, Leroy E. (ed.), *G.W. Leibniz: Philosophical Papers and Letters* (Dordrecht: Kluwer Academic Publishers, 1989).

SECONDARY SOURCES

Adams, Robert M. *Leibniz: Determinist, Theist, Idealist* (New York: Oxford University Press, 1994).
Aiton, E.J. *Leibniz: A Biography* (Bristol & Boston: Adams Hilger, 1985).
Alexander, H. G. *The Leibniz-Clarke Correspondence* (New York:

Manchester University Press, 1970).
Andrade, E.N. da C. *Sir Isaac Newton* (London: Collins, 1954).
Arthur, Richard T.W. "Leibniz's Theory of Time," K. Okruhlik and J.R. Brown (eds.), *The Natural Philosophy of Leibniz* (Reidel, 1985), pp. 263-313.
– "Space and Relativity in Newton and Leibniz," *British Journal of Philosophy of Science*, vol. 45 (1994), pp. 219-40.
– "Newton's Fluxions and Equably Flowing Time," *Studies in History and Philosophy of Science*, vol. 26 (1995), pp. 323-351.
Axtell, James L. "Locke's Review of the *Principia*," *Notes and Records of the Royal Society of London*, vol. 2 (1965), pp. 152-61.
– "Locke, Newton and the Two Cultures", John Yolton (ed.), *John Locke: Problems and Perspectives* (Cambridge: Cambridge University Press, 1969), pp. 172-181.
Ball, Bryan W. *A Great Expectation: Eschatological Thought in English Protestantism to 1660* (Leiden, 1975).
Baron, M.E. *The Origins of the Infinitesimal Calculus* (Oxford: Pergamon Press, 1969).
Bechler, Zev (ed.), *Contemporary Newtonian Research* (Dordrecht: D. Reidel Publishing Company, 1982).
Ben-Chaim, Michael. "The Empiric Experience and the Practice of Autonomy," *Studies in History and Philosophy of Science*, vol. 23, (1992), pp. 533-555.
– "Doctrine and Use: Newton's 'Gift of Preaching,'" *History of science*, vol. 36 (1998), pp. 269-298.
Bertoloni-Meli, Domenico. *Equivalence and Priority: Newton versus Leibniz* (Oxford: Clarendon Press, 1993).
Blumenfeld, David. "Leibniz's Theory of the Striving Possibilities," *Studia Leibnitiana*, vol. 5, no. 2 (1973), pp. 163-77.
Bos, H.J.M. "Differentials, Higher-Order Differentials and the Derivative in the Leibnizian Calculus," *Archive for History of Exact Sciences*, vol. 14 (1974), pp. 1-90.
– "Fundamental Concepts of the Leibnizian Calculus," *Studia Leibnitianna* Sonderheft 14, (1986).
Boyer, C. *History of the Calculus* (New York: Dover, 1949).
Brackenridge, J. Bruce. "The Critical Role of Curvature in Newton's Developing Dynamics," P.M. Harman and Alan E. Shapiro (eds.), *The Investigation of Difficult Things* (Cambridge: Cambridge University Press, 1992), pp. 231-260.
Brandom, Robert B. "Leibniz and Degrees of Perception," *Journal of the History of Philosophy*, vol. 19 (1981), pp. 447-479.
Brewster, Sir David. *Memoirs of the Life, Writings, and Discoveries of Sir Isaac Newton.* (2 vol's; Edinburgh, 1850).
Cajori, Florian. *A History of Mathematics* (New York: Macmillan, 1919).

Cassirer, Ernest. "Newton and Leibniz," *Philosophical Review*, vol. 52 (1943), pp. 366-91.

Castillejo, David. "A Theory of Shifting Relationships in Knowledge as Seen in Medieval and Modern Times with a Reconstruction of Newton's Thought and Essays on Patronage," Ph.D. thesis, 3 vols. Cambridge University, 1967.

– *A Report on the Yahuda Collection of Newton MSS. Bequeathed to the Jewish National and University Library at Jerusalem* (Jerusalem, 1969).

– *The Expanding Force in Newton's Cosmos*, (Madrid, 1981).

Christianson, Paul. *Reformers and Babylon: English Apocalyptic Visions from the Reformation to the Eve of the Civil War* (Toronto, 1978).

Cohen, I.B. "Isaac Newton's *Principia*, the Scriptures, and the divine providence," Sidney Morgenbesser, et al. (ed.), *Philosophy, Science and Medicine* (New York: St. Martin's Press, 1969), pp. 523-48.

– "Newton's Second Law and the Concept of Force in the *Principia*," Robert Palter (ed.), *The Annus Mirabilis of Sir Isaac Newton* (Cambridge: MIT Press, 1970), pp. 186-91.

– *Introduction to Newton's Principia* (Cambridge, 1971).

– "The *Principia* Universal Gravitation, and 'Newtonian Style'" Zev Bechler (ed.), *Contemporary Newtonian Research* (Dordrecht: D. Reidel Publishing Company, 1982), pp. 21-109.

Copenhaver, B.P. "Jewish Theologies of Space in the Scientific Revolution: Henry More, Joseph Raphson, Isaac Newton and their Predecessors," *Annals of Science*, vol. 37 (1980), pp. 489-548.

Coudert, Allison. "The *Kabbala Denudata*: Converting Jews or Seducing Christians?" R. H. Popkin and G. M. Weiner (eds.), *Jewish Christians and Christian Jews from the Renaissance to the Enlightenment* (Dordrecht: Kluwer, 1994).

– *Leibniz and the Kabbalah* (Dordrecht: Kluwer, 1995).

Cover, J.A. and Hartz, Glenn A. "Space and Time in the Leibnizian Metaphysic," *Nous*, vol. 22 (1988), pp. 493-519.

Crombie, A.C. *Robert Groseteste and the Origins of Experimental Science, 1100-1700* (Oxford: The Clarendon Press, 1953).

De Gandt, François. *Force and Geometry in Newton's Principia* (Princeton: Princeton University Press, 1995).

De Smet, Rudolf and Verelst, Karin. "Newton's Scholium Generale: the Platonic and Stoic legacy — Philo, Justus Lipsius and the Cambridge Platonists," *History of Science* vol. 39 (2001), pp. 1-30.

Dewey, John. "Leibniz's New Essays Concerning the Human

Understanding," *John Dewey: The Early Works, 1882-88* (London and Amsterdam: Southern Illinois University Press, 1975), pp. 342-73.

Dobbs, B.J.T. *The Foundations of Newton's Alchemy, or 'The Hunting of the Greene Lyon'* (Cambridge: Cambridge University Press, 1975).

– "Newton's *Commentary* on the *Emerald Tablet* of Hermes Trismegistus: Its Scientific and Theological Significance," I. Merkel and A.G. Debus, (eds.), *Hermeticism and The Renaissance*, (Folger Books, 1988), pp. 183-91.

– "Newton as Alchemist and Theologian," N.J.W. Thrower (ed.), *Standing on the Shoulders of Giants* (Berkeley, 1990), pp. 129-140.

– *The Janus Face of Genius: The Role of Alchemy in Newton's Thought* (Cambridge: Cambridge University Press, 1991).

Faur, José. "Newton, Maimonides, and Esoteric Knowledge," *Cross Currents: Religion & Intellectual Life*, vol. 40 (1990), pp. 526-540.

Figala Karin. "Newton as Alchemist," *History of Science*, vol. 15 (1977), pp. 102-37.

Firth, Katherine R. *The Apocalyptic Tradition in Reformation Britain 1530-1645* (Oxford, 1979).

Force, James E., "Hume and the Relation of Science to religion among certain members of the Royal Society," *Journal of the History of Ideas*, vol. 45, (1984).

– *William Whiston: Honest Newtonian* (Cambridge, 1985).

– and Popkin, Richard H. (eds.), *Essays on the Context, Nature, and Influence of Isaac Newton's Theology* (Dordrecht: Kluwer Academic Publishers, 1990).

– "Newton's God of Dominion: The Unity of Newton's Theological, Scientific, and Political Thought," James E. Force and Richard H. Popkin (eds.), *Essays on the Context, Nature, and Influence of Isaac Newton's Theology* (Dordrecht, 1990), pp. 75-102.

– "Newton's 'Sleeping Argument' and the Newtonian Synthesis of Science and Religion," N.J.W. Thrower (ed.), *Standing on the Shoulders of Giants* (Berkeley, 1990).

– "The Newtonians and Deism," Force and Popkin (eds.), *Essays on the Context, Nature, and Influence of Isaac Newton's Theology* (Dordrecht, 1990), pp. 53-66.

– and Popkin, Richard (eds.), *The Books of Nature and Scripture: Recent Essays on Natural Philosophy, Theology, and Biblical Criticism in the Netherlands of Spinoza's Time and the British Isles of Newton's Time* (Dordrecht, 1994).

– "Newton, the Lord God of Israel and Knowledge of Nature," Richard H. Popkin and G.M. Weiner (eds.), *Jewish*

Christians and Christian Jews: From the Renaissance to the Enlightenment (Dordrecht, 1994), 131-58.

– "Samuel Clarke's Four Categories of Deism, Isaac Newton and the Bible," R. Popkin (ed.), *Skepticism in the History of Philosophy*, (Dordrecht, 1996), pp. 53-74.

– and Popkin, Richard (eds.), *Newton and Religion: Context, Nature, and Influence* (Dordrecht: Kluwer, 1999).

– "Newton, the 'Ancients,' and the 'Moderns,'" Force and Popkin (eds.), *Newton and Religion: Context, Nature, and Influence* (Dordrecht, 1999), pp. 237-59.

– "The Nature of Newton's 'Holy Alliance' between Science and Religion: From the Scientific Revolution to Newton (and back again)," Margaret J. Osler (ed.), *Rethinking the Scientific Revolution* (Cambridge, 2000), pp. 247-70.

Freudenthal, Gideon. *Atom and Individual in the Age of Newton: On the Genesis of the Mechanistic World View* (Dordrecht: Reidel, 1986).

Funkenstein, Amos. *Theology and the Scientific Imagination* (Princeton: Princeton University Press, 1986).

Gabbey, Alan. "Force and Inertia in Seventeenth-Century Dynamics," *Studies in History and Philosophy of Science*, vol. 2 (1971), pp. 1-67.

Garber, Daniel. "Leibniz: Physics and Philosophy," Nicholas Jolley (ed.), *The Cambridge Companion to Leibniz* (Cambridge: Cambridge University Press, 1995), pp. 270-353.

Goldish, Matt. "Newton on Kabbalah," Force and Popkin (eds.), *The Books of Nature and Scripture: Recent Essays on Natural Philosophy, Theology, and Biblical Criticism in the Netherlands of Spinoza's Time and the British Isles of Newton's Time* (Dordrecht, 1994), pp. 89-103.

– *Judaism in the Theology of Sir Isaac Newton* (Dordrecht: Kluwer, 1998).

Gouk, Penelope. "The Harmonic Roots of Newtonian Science," J.Fauvel, Raymond Flood, Michael Shortland and Robert Wilson (eds.), *Let Newton be! A New Perspective on his Life and Works*, (Oxford, 1988), pp. 101-125.

Gregory, David. *David Gregory, Isaac Newton and their Circle*, W.G. Hiscock (ed.), (Oxford, 1937).

Guicciardini, Niccolo. *Reading the* ***Principia****: The Debate on Newton's Mathematical Methods for Natural Philosophy from 1687 to 1736* (Cambridge: Cambridge University Press, 1999).

Haase, Carl and Totok, Wilhelm (eds.), *Leibniz: sein Leben - sein Werken - seine Welt* (Hannover: Verlag fur Literatur und Zeitgeschehen, 1966).

Hall, Rupert *Philosophers at War: The Quarrel between Newton and*

Leibniz (Cambridge: Cambridge University Press, 1980).

– "Newton and the Absolutes: Sources," P.M. Harman and Alan E. Shapiro (eds.), *The Investigation of Difficult Things: Essays on Newton and the Exact Sciences in Honour of D.T. Whiteside* (Cambridge: Cambridge University Press, 1992), pp. 261-287.

Harman, P.M. and Shapiro, Alan E. (eds.), *The Investigation of Difficult Things: Essays on Newton and the Exact Sciences in Honour of D.T. Whiteside* (Cambridge: Cambridge University Press, 1992).

Henry, John. "'Pray not Ascribe that Notion to Me': God and Newton's Gravity," James and Popkin (eds.), *The Books of Nature and Scripture* (Dordrecht, 1994), pp. 123-47.

Herivel, John (ed.), *The Background to* Newton's *Principia* (Oxford: Oxford Clarendon Press, 1996).

Hessen, Borris. *The Social Economic Roots of Newton's '**Principia**'* (New York: Howard Fertig, 1971).

Heyd, Michael. "*Be Sober and Reasonable*:" *The Critique of Enthusiasm in the Seventeenth and Early Eighteenth Centuries*, (Leiden: Brill, 1995).

Hoffman, J.E. *Leibniz in Paris 1672-76: His Growth to Mathematical Maturity* (Cambridge: Cambridge University Press, 1974).

Horvath, Miklos. "On the Attempts Made by Leibniz to Justify his Calculus," *Studia Leibnitia*, vol. 25 (1986), pp. 60-71.

Hoskin, M. A. "Newton's Providence and the Universe of Stars," *Journal of the History of Astronomy*, vol. 8 (1977), pp. 77-101.

Hunter, Michael. *Science and Society in Restoration England* (Cambridge: Cambridge University Press, 1981).

Hutton, Sarah. "More, Newton, and the Language of Biblical Prophecy," Force and Popkin (eds.), *The Books of Nature and Scripture: Recent Essays on Natural Philosophy, Theology, and Biblical Criticism in the Netherlands of Spinoza's Time and the British Isles of Newton's Time* (Dordrecht, 1994), pp. 39-53.

– "The Seven Trumpets and the Seven Vials: Apocalypticism and Christology in Newton's Theological Writings," Force and Popkin (eds.), *Newton and Religion: Context, Nature, and Influence* (Dordrecht: Kluwer, 1999), pp. 165-178.

Iliffe, Robert."'In the Warehouse': Privacy, Property and Priority in the Early Royal Society," *History of Science*, vol. 30 (1992), pp. 29-68.

– "'Making a Shew': Apocalyptic Hermeneutics and the Sociology of Christian Idolatry in the Work of Isaac Newton and Henry More," Force and Popkin (eds.), *The Books of Nature and Scripture*, (Kluwer, 1994), pp. 55-88.

– "'That Puzzling Problem': Isaac Newton and the Political Philosophy of Self," *Medical History*, vol. 39 (1995), pp. 433-458.

– "'Is he like other men?' The meaning of the *Principia Mathematica*, and the author as idol," Gerald Maclean (ed.), *Culture and Society in the Stuart Restoration*, (Cambridge, 1995).

– "Those 'Whose Business It Is To Cavill': Newton's Anti-Catholicism," Force and Popkin (eds.), Force and Popkin (eds.), *Newton and Religion: Context, Nature, and Influence* (Dordrecht: Kluwer, 1999), pp. 97-121.

Jacob, Margaret. *The Newtonian in the English Revolution, 1689-1720* (Ithaca: Cornell University Press, 1976).

– and Dobbs, B. J.T. *Newton and the Culture of Newtonianism* (New Jersey, Humanity Press, 1995).

James, Jamie. *The Music of the Spheres: Music, Science and the Natural Order of the Universe* (New York: Copernicus, 1993).

Jesseph, Douglas M. "Leibnitz on the Foundation of the Calculus: The Question of the Reality of Infinitesimal Magnitudes," *Perspective on Science*, vol. 6 (1998), pp. 6-40.

Jolley, Nicholas (ed.), *The Cambridge Companion to Leibniz* (4 vol's; Cambridge: Cambridge University Press, 1995).

– *Leibniz and Locke* (Oxford: Oxford University Press, 1984).

Kerszberg, Pierre. "The Cosmological Question in Newton's Science," *Osiris*, vol. 2 (1986), pp. 69-106.

Kitcher, Philip. "Fluxions, Limits, and Infinite Littleness: A Study of Newton's Presentation of the Calculus," *Isis*, vol. 64 (1973), pp. 33-49.

Kline, Morris. *Mathematical Thought from Ancient to Modern Times* (New York: Oxford University Press, 1972).

Kochavi, Matania Z. "One Prophet Interprets Another: Sir Isaac Newton and Daniel," R. Popkin and J. Force (eds.), *The Books of Nature and Scripture* (Dordrecht, 1994), pp.105-122.

Knoespel, Kenneth J. "Newton in the School of Time: The *Chronology of Ancient Kingdoms Amended* and the Crisis of Seventeenth-Century Historiography," *The Eighteenth Century*, vol. 30, no.3 (1989), pp. 19-41.

– "Interpretive Strategies in Newton's *Theologiae Gentilis Origines Philosophiae*," Force and Popkin (eds.), *Newton and Religion: Context, Nature, and Influence* (Dordrecht, 1999), pp. 179-202.

Kubrin, David. "Newton and the Cyclical Cosmic: Providence and the Mechanical Philosophy," *Journal of the History of Ideas*, vol. 28 (1967), pp. 324-346.

Kulstad, Mark. *Leibniz on Apperception, Consciousness, and Reflection* (Munich: Philosophia, 1991).

Linde, A. "The Self Reproducing Inflationary Universe," *Scientific American*, vol. 271, no. 5 (1994).

Locke, John. *An Essay Concerning Human Understanding* A. D. Woozley (ed.), (New York: New American Library, 1974.)

Mamiani, Maurizio. "The Rhetoric of Certainty: Newton's Method in Science and in the Interpretation of the Apocalypse," M. Pera and W.R. Shea (eds.), *Persuading Science: The Art of Scientific Rhetoric*, (Canton, 1991), pp. 157-172.

– "To Twist the Meaning: Newton's *Regulae Philosophandi* Revisited," *Isaac Newton's Natural Philosophy*, Jed Z. Buchwald and I. Bernard Cohen (eds.), (Cambridge, 2001), pp. 3-14.

Mandelbrote, Scott. "'A Duty of the Greatest Moment': Isaac Newton and the Writing of Biblical Criticism," *British Journal for History of Science*, vol. 26 (1993), pp. 281-302.

– "Isaac Newton and Thomas Burnet: Biblical Criticism and the Crisis of Late Seventeenth-Century England," Force and Popkin (eds.), *The Books of Nature and Scripture: Recent Essays on Natural Philosophy, Theology, and Biblical Criticism in the Netherlands of Spinoza's Time and the British Isles of Newton's Time* (Dordrecht, 1994), pp. 149-178.

Manuel, Frank. *Isaac Newton, Historian* (Cambridge, Mass., 1963).

– *A Portrait of Isaac Newton* (New York: Da Capo Press, 1968).

– *The Religion of Isaac Newton* (Oxford: Clarendon Press, 1974).

Markley, Robert. "Newton, Corruption, and the Tradition of Universal History," J.E. Force and R.H. Popkin (eds.), *Newton and Religion: Context, Nature and Influence*, (Dordrecht, 1999), pp. 121-145.

McGuire J.E. and Rattansi, P.M. "Newton and the Pipes of Pan," *Notes and Records of the Royal Society of London*, vol. 21 (1966), pp. 108-143.

McGuire, J. E. "Force, Active Principles, and Newton's Invisible Realm," *Ambix*, vol. 25 (1968), pp. 154-208.

– "The Origin of Newton's Doctrine of Essential Qualities," *Centaurus*, vol. 12 (1968), pp. 238-9.

– "Force, Active Principles, and Newton's Invisible Realm," *Ambix*, vol. 25 (1968), pp. 154-208.

– "Comment on Cohen," Robert Palter (ed.), *The Annus Mirabilis of Sir Isaac Newton: 1666-1966* (Cambridge: MIT Press, 1970), pp. 143-185.

– "Neoplatonism and Active Principles: Newton and the *Corpus Hermeticum*, in R.S.Westman and J.E. McGuire (eds.), *Hermeticism and the Scientific Revolution* (Los Angeles,

1977).
– "Newton on Place, Time and God: An Unpublished Source," *The British Journal for the History of Science*, vol. 11 (1978), pp. 114-129.
– "Existence, Actuality and Necessity: Newton on Space and Time," *Annals of Science*, vol. 35 (1978), pp. 463-508.
– "Space, Infinity, and Indivisibilty: Newton on the Creation of Matter," Zev Bechler (ed.), *Contemporary Newtonian Research* (Dordrecht: Reidel, 1982).
– "The Fate of the Date: The Theology of Newton's *Principia* Revisited," Margaret J. Osler (ed.), *Rethinking the Scientific Revolution*, (Cambridge, 2000), pp. 271-295.
McMullin, Ernan. *Newton on Matter and Activity* (Notre Dame, Indiana: University of Notre Dame Press, 1978).
McRae, Robert. *Leibniz: Perception, Apperception and Thought* (Toronto: Toronto University Press, 1976).
Meyer, R.W. *Leibnitz and the Seventeenth-Century Revolution* (Cambridge: Bowes & Bowes, 1952).
Miller, K. *Leibniz – Bibliographie* (Frankfurt am Main: Klostermann, 1984).
More, Louis T. *Isaac Newton: A Biography* (New York: Dover Publications, 1934).
Northrop, F.S.C. "Leibniz's Theory of Space," *Journal of the History of Ideas*, vol. 8 (1946), pp. 422-446.
Palter, Robert (ed.), *The Annus Mirabilis of Sir Isaac Newton: 1666-1966* (Cambridge: MIT Press, 1970).
Patai, Raphael. *Man and Temple: In Ancient Jewish Myth and Ritual.* (New York: Ktav Publishing House, 1947).
Popkin, Richard. *The History of Skepticism from Erasmus to Spinoza* (Berkeley & Los Angeles: University of California Press, 1964).
– (ed.), *Millenarianism and Messianism in English Literature and Thought 1650-1800* (Leiden, 1988).
– "Newton's Biblical Theology and his Theological Physics," P.B. Scheurer & G. Debrock (eds.), *Newton's Scientific and Philosophical Legacy*, (Dordrecht, 1988).
– "Newton and Maimonides," Ruth Link-Salinger (ed.), *A Straight Path: Studies in Medieval Philosophy and Culture.* (Washington, D.C.: Catholic University of America Press, 1988), pp. 216-229
– "Newton as a Bible Scholar," J.E. Force and R.H. Popkin (eds.), *Essays on the Context, Nature, and Influence of Isaac Newton's Theology.* (Dordrecht: 1990), pp. 103-118.
– and G. M. Weiner (eds.), *Jewish Christians and Christian Jews from the Renaissance to the Enlightenment.* (Dordrecht: Kluwer, 1994).

Ramati, Ayval. "Harmony at a Distance: Leibniz's Scientific Academies," *Isis*, vol. 87 (1996), pp. 430-452.
– "The Hidden Truth of Creation: Newton's Method of Fluxions," *The British Journal for the History of Science*, vol. 34 (2001), pp. 417-438.
Rattansi, P.M, and McGuire J.E. "Newton and the Pipes of Pan," *Notes and Records of the Royal Society of London*, vol. 21 (1966), pp. 108-143.
– "Newton's Alchemical Studies," Allen G. Debus (ed.), *Science, Medicine and Society in the Renaissance*, vol. 2, (New York, 1972), pp. 167-82.
– "Newton and the Wisdom of the Ancients," J.Fauvel, Raymond Flood, Michael Shortland and Robert Wilson (eds.), *Let Newton be! A New Perspective on his Life and Works*, (Oxford, 1988), pp.185-201
Rescher, Nicholas. *The Philosophy of Leibniz* (Englewood Cliffs: Prentice Hall, 1967).
Rogers, G.A.J. "Locke's Essay and Newton's *Principia*," *Journal of the History of Ideas*, vol. 37 (1978), pp. 121-136.
Rutherford, Donald. *Leibniz and the Rational Order* (Cambridge: Cambridge University Press, 1995).
Schaffer, Simon. "Glass works: Newton's prisms and the uses of experiment," D. Gooding, T. Pinch, S. Schaffer (eds.), *The Uses of Experiment* (Cambridge: Cambridge University Press, 1989), pp. 67-104.
– "Comets & Idols: Newton's Cosmology and Political Theology," P. Theerman and A. Seeff (eds.) *Action and Reaction: Proceedings of a Symposium to Commemorate the Tercentenary of Newton's* ***Principia***, (London, 1992), pp. 206-231.
Schechner Genuth, Sara "Newton and the Ongoing Teleological Role of Comets," N.J.W. Thrower (ed.), *Standing on the Shoulders of Giants* (Berkeley, 1990), pp. 299-311.
– *Comets, Popular Culture and the Birth of Modern Cosmology* (Princeton, 1997).
Shapin, Steven. "On Gods and Kings: Natural Philosophy and Politics in the Leibniz-Clarke Correspondence," *Isis*, vol. 72 (1981), pp. 187-215.
Shoshani, Yakir. *Thoughts on Reality* (Hebrew), (Tel-Aviv: Ministry of Defence, Israel, 1999) pp. 62-70.
Snobelen, Stephen D. "Isaac Newton, Heretic: the Strategies of a Nicodemite," *The British Journal for the History of Science*, vol. 32 (1999), pp. 381-419.
– "William Whiston: Natural Philosopher, Prophet, Primitive Christian," Ph.D thesis, University of Cambridge, 2000.
– "'God of gods, and Lord of lords': the Theology of Isaac Newton's

General Scholium to the *Principia*," *Osiris*, vol. 16 (2001), pp. 169-208.
– "'The Mystery of this Restitution of all Things': Isaac Newton on the Return of the Jews," James E. Force and Richard H. Popkin (eds.), *The Millenarian Turn: Millenarian Contexts of Science, Politics, and Everyday Anglo-American Life in the Seventeenth and Eighteenth Centuries*, (Dordrecht, 2001), pp. 95-118.
– "Mathematicians, Historians and Newton's *Principia*," *Annals of Science*, vol. 58 (January 2001), pp. 75-84.
Stewart, Larry. "Seeing Through the Scholium: Religion and Reading Newton in the Eighteenth Century," *History of Science*, vol. 34 (1996), pp. 123-65.
Tammy, Martin. "Newton, Creation, and Perception," *Isis*, vol. 70 (1979), pp. 48-58.
Vailati, Ezio. "Leibniz and Clarke on Miracles," *Journal of the History of Philosophy*, vol. 33 (1995), pp. 563-591.
– *Leibniz and Clarke. A Study of their Correspondence* (New York, Oxford University Press, 1997).
Verelst, Karin and De Smet, Rudolf. "Newton's Scholium Generale: the Platonic and Stoic legacy — Philo, Justus Lipsius and the Cambridge Platonists," *History of Science* vol. 39 (2001), pp. 1-30.
Westfall, Richard S. "The Development of Newton's Theory of Colour," *Isis*, vol. 53 (1962), pp. 339-358.
– "Newton's Reply to Hooke and the Theory of Colors," *Isis*, vol. 34 (1963).
– *Never at Rest* (Cambridge: Cambridge University Press, 1980)
– "Newton's Theological Manuscripts," Z. Bechler (ed.), *Contemporary Newtonian Research*, (Reidel, 1982).
– "Isaac Newton's *Theologiae Gentilis Origines Philosophicae*," W. Warren (ed.), *The Secular Mind: Transformations of Faith in Modern Europe- Essays Presented to Franklin L. Baumenr, Randolf W. Townsend Professor of History, Yale University* (New York, 1982), pp. 15-34.
– *The Life of Isaac Newton* (New York: Cambridge University Press, 1993).
Whiston, William. *A New Theory of the Earth* (London, 1696).
– *A collection of Authentick Records Belonging to the Old and New Testament* (London, 1727).
– *Memoirs of the Life and Writings of Mr. William Whiston. Containing Memoirs of Several of his Friends*, (2vols., London, 1753)
White, Michael. *Isaac Newton: The Last Sorcerer* (London, 1998).
Whiteside, D.T. "Patterns of Mathematical Thought in the Late Seventeenth Century," *Archive for History of Exact Science*,

vol. 1 (1961), pp. 179-338.

– "Isaac Newton: Birth of a Mathematician," *Notes and Records of the Royal Society*, vol. 19 (1964), pp. 53-62.

Winterbourne, A.T. "On the Metaphysics of Leibnizian Space and Time," *Studies in the History and Philosophy of Science*, vol. 13 (1982), pp. 201-14.

INDEX

Y

Z

ARCHIVES INTERNATIONALES D'HISTOIRE DES IDÉES
*
INTERNATIONAL ARCHIVES OF THE HISTORY OF IDEAS

---

1. E. Labrousse: *Pierre Bayle.* Tome I: *Du pays de foix à la cité d'Erasme.* 1963; 2nd printing 1984 ISBN 90-247-3136-4
   For Tome II *see below under Volume 6.*
2. P. Merlan: *Monopsychism, Mysticism, Metaconsciousness.* Problems of the Soul in the Neoaristotelian and Neoplatonic Tradition. 1963; 2nd printing 1969 ISBN 90-247-0178-3
3. H.G. van Leeuwen: *The Problem of Certainty in English Thought, 1630–1690.* With a Preface by R.H. Popkin. 1963; 2nd printing 1970 ISBN 90-247-0179-1
4. P.W. Janssen: *Les origines de la réforme des Carmes en France au 17e Siècle.* 1963; 2nd printing 1969 ISBN 90-247-0180-5
5. G. Sebba: *Bibliographia Cartesiana.* A Critical Guide to the Descartes Literature (1800–1960). 1964 ISBN 90-247-0181-3
6. E. Labrousse: *Pierre Bayle.* Tome II: *Heterodoxie et rigorisme.* 1964 ISBN 90-247-0182-1
7. K.W. Swart: *The Sense of Decadence in 19th-Century France.* 1964 ISBN 90-247-0183-X
8. W. Rex: *Essays on Pierre Bayle and Religious Controversy.* 1965 ISBN 90-247-0184-8
9. E. Heier: *L.H. Nicolay (1737–1820) and His Contemporaries.* Diderot, Rousseau, Voltaire, Gluck, Metastasio, Galiani, D'Escherny, Gessner, Bodmer, Lavater, Wieland, Frederick II, Falconet, W. Robertson, Paul I, Cagliostro, Gellert, Winckelmann, Poinsinet, Lloyd, Sanchez, Masson, and Others. 1965 ISBN 90-247-0185-6
10. H.M. Bracken: *The Early Reception of Berkeley's Immaterialism, 1710–1733.* [1958] Rev. ed. 1965 ISBN 90-247-0186-4
11. R.A. Watson: *The Downfall of Cartesianism, 1673–1712.* A Study of Epistemological Issues in Late 17th-Century Cartesianism. 1966 ISBN 90-247-0187-2
12. R. Descartes: *Regulæ ad Directionem Ingenii.* Texte critique établi par Giovanni Crapulli avec la version hollandaise du 17e siècle. 1966 ISBN 90-247-0188-0
13. J. Chapelain: *Soixante-dix-sept Lettres inédites à Nicolas Heinsius (1649–1658).* Publiées d'après le manuscrit de Leyde avec une introduction et des notes par B. Bray. 1966 ISBN 90-247-0189-9
14. C. B. Brush: *Montaigne and Bayle.* Variations on the Theme of Skepticism. 1966 ISBN 90-247-0190-2
15. B. Neveu: *Un historien à l'Ecole de Port-Royal.* Sébastien le Nain de Tillemont (1637–1698). 1966 ISBN 90-247-0191-0
16. A. Faivre: *Kirchberger et l'Illuminisme du 18e siècle.* 1966 ISBN 90-247-0192-9
17. J.A. Clarke: *Huguenot Warrior.* The Life and Times of Henri de Rohan (1579–1638). 1966 ISBN 90-247-0193-7
18. S. Kinser: *The Works of Jacques-Auguste de Thou.* 1966 ISBN 90-247-0194-5
19. E.F. Hirsch: *Damião de Gois.* The Life and Thought of a Portuguese Humanist (1502–1574). 1967 ISBN 90-247-0195-3
20. P.J.S. Whitemore: *The Order of Minims in 17th-Century France.* 1967 ISBN 90-247-0196-1
21. H. Hillenaar: *Fénelon et les Jésuites.* 1967 ISBN 90-247-0197-X
22. W.N. Hargreaves-Mawdsley: *The English Della Cruscans and Their Time, 1783–1828.* 1967 ISBN 90-247-0198-8
23. C.B. Schmitt: *Gianfrancesco Pico della Mirandola (1469–1533) and his Critique of Aristotle.* 1967 ISBN 90-247-0199-6
24. H.B. White: *Peace among the Willows.* The Political Philosophy of Francis Bacon. 1968 ISBN 90-247-0200-3

ARCHIVES INTERNATIONALES D'HISTOIRE DES IDÉES
*
INTERNATIONAL ARCHIVES OF THE HISTORY OF IDEAS

25. L. Apt: *Louis-Philippe de Ségur.* An Intellectual in a Revolutionary Age. 1969 ISBN 90-247-0201-1
26. E.H. Kadler: *Literary Figures in French Drama (1784–1834).* 1969 ISBN 90-247-0202-X
27. G. Postel: *Le Thrésor des prophéties de l'univers.* Manuscrit publié avec une introduction et des notes par F. Secret. 1969 ISBN 90-247-0203-8
28. E.G. Boscherini: *Lexicon Spinozanum.* 2 vols., 1970 Set ISBN 90-247-0205-4
29. C.A. Bolton: *Church Reform in 18th-Century Italy.* The Synod of Pistoia (1786). 1969 ISBN 90-247-0208-9
30. D. Janicaud: *Une généalogie du spiritualisme français.* Aux sources du bergsonisme: [Félix] Ravaisson [1813–1900] et la métaphysique. 1969 ISBN 90-247-0209-7
31. J.-E. d'Angers: *L'Humanisme chrétien au 17ᵉ siècle.* St. François de Sales et Yves de Paris. 1970 ISBN 90-247-0210-0
32. H.B. White: *Copp'd Hills towards Heaven.* Shakespeare and the Classical Polity. 1970 ISBN 90-247-0250-X
33. P.J. Olscamp: *The Moral Philosophy of George Berkeley.* 1970 ISBN 90-247-0303-4
34. C.G. Noreña: *Juan Luis Vives (1492–1540).* 1970 ISBN 90-247-5008-3
35. J. O'Higgens: *Anthony Collins (1676–1729), the Man and His World.* 1970 ISBN 90-247-5007-5
36. F.T. Brechka: *Gerard van Swieten and His World (1700–1772).* 1970 ISBN 90-247-5009-1
37. M.H. Waddicor: *Montesquieu and the Pilosophy of Natural Law.* 1970 ISBN 90-247-5039-3
38. O.R. Bloch: *La Philosophie de Gassendi (1592–1655).* Nominalisme, matérialisme et métaphysique. 1971 ISBN 90-247-5035-0
39. J. Hoyles: *The Waning of the Renaissance (1640–1740).* Studies in the Thought and Poetry of Henry More, John Norris and Isaac Watts. 1971 ISBN 90-247-5077-6
    For Henry More, *see also below under Volume 122 and 127.*
40. H. Bots: *Correspondance de Jacques Dupuy et de Nicolas Heinsius (1646–1656).* 1971 ISBN 90-247-5092-X
41. W.C. Lehmann: *Henry Home, Lord Kames, and the Scottish Enlightenment.* A Study in National Character and in the History of Ideas. 1971 ISBN 90-247-5018-0
42. C. Kramer: *Emmery de Lyere et Marnix de Sainte Aldegonde.* Un admirateur de Sébastien Franck et de Montaigne aux prises avec le champion des calvinistes néerlandais.[Avec le texte d'Emmery de Lyere:] *Antidote ou contrepoison contre les conseils sanguinaires et envinemez de Philippe de Marnix Sr. de Ste. Aldegonde.* 1971 ISBN 90-247-5136-5
43. P. Dibon: *Inventaire de la correspondance (1595–1650) d'André Rivet (1572–1651).* 1971 ISBN 90-247-5112-8
44. K.A. Kottman: *Law and Apocalypse.* The Moral Thought of Luis de Leon (1527?–1591). 1972 ISBN 90-247-1183-5
45. F.G. Nauen: *Revolution, Idealism and Human Freedom.* Schelling, Hölderlin and Hegel, and the Crisis of Early German Idealism. 1971 ISBN 90-247-5117-9
46. H. Jensen: *Motivation and the Moral Sense in Francis Hutcheson's* [1694–1746] *Ethical Theory.* 1971 ISBN 90-247-1187-8
47. A. Rosenberg: *[Simon] Tyssot de Patot and His Work (1655–1738).* 1972 ISBN 90-247-1199-1
48. C. Walton: *De la recherche du bien.* A study of [Nicolas de] Malebranche's [1638–1715] Science of Ethics. 1972 ISBN 90-247-1205-X

ARCHIVES INTERNATIONALES D'HISTOIRE DES IDÉES
*
INTERNATIONAL ARCHIVES OF THE HISTORY OF IDEAS

49. P.J.S. Whitmore (ed.): *A 17th-Century Exposure of Superstition.* Select Text of Claude Pithoys (1587–1676). 1972 ISBN 90-247-1298-X
50. A. Sauvy: *Livres saisis à Paris entre 1678 et 1701.* D'après une étude préliminaire de Motoko Ninomiya. 1972 ISBN 90-247-1347-1
51. W.R. Redmond: *Bibliography of the Philosophy in the Iberian Colonies of America.* 1972 ISBN 90-247-1190-8
52. C.B. Schmitt: *Cicero Scepticus.* A Study of the Influence of the *Academica* in the Renaissance. 1972 ISBN 90-247-1299-8
53. J. Hoyles: *The Edges of Augustanism.* The Aesthetics of Spirituality in Thomas Ken, John Byrom and William Law. 1972 ISBN 90-247-1317-X
54. J. Bruggeman and A.J. van de Ven (éds.): *Inventaire* des pièces d'Archives françaises se rapportant à l'Abbaye de Port-Royal des Champs et son cercle et à la Résistance contre la Bulle *Unigenitus* et à l'Appel. 1972 ISBN 90-247-5122-5
55. J.W. Montgomery: *Cross and Crucible.* Johann Valentin Andreae (1586–1654), Phoenix of the Theologians. Volume I: Andreae's Life, World-View, and Relations with Rosicrucianism and Alchemy; Volume II: The *Chymische Hochzeit* with Notes and Commentary. 1973 Set ISBN 90-247-5054-7
56. O. Lutaud: *Des révolutions d'Angleterre à la Révolution française.* Le tyrannicide & *Killing No Murder* (Cromwell, *Athalie*, Bonaparte). 1973 ISBN 90-247-1509-1
57. F. Duchesneau: *L'Empirisme de Locke.* 1973 ISBN 90-247-1349-8
58. R. Simon (éd.): *Henry de Boulainviller* – Œuvres Philosophiques, Tome I. 1973 ISBN 90-247-1332-3

    For Œuvres Philosophiques, Tome II *see below under Volume 70.*
59. E.E. Harris: *Salvation from Despair.* A Reappraisal of Spinoza's Philosophy. 1973 ISBN 90-247-5158-6
60. J.-F. Battail: *L'Avocat philosophe Géraud de Cordemoy (1626–1684).* 1973 ISBN 90-247-1542-3
61. T. Liu: *Discord in Zion.* The Puritan Divines and the Puritan Revolution (1640–1660). 1973 ISBN 90-247-5156-X
62. A. Strugnell: *Diderot's Politics.* A Study of the Evolution of Diderot's Political Thought after the *Encyclopédie.* 1973 ISBN 90-247-1540-7
63. G. Defaux: *Pantagruel et les Sophistes.* Contribution à l'histoire de l'humanisme chrétien au 16e siècle. 1973 ISBN 90-247-1566-0
64. G. Planty-Bonjour: *Hegel et la pensée philosophique en Russie (1830–1917).* 1974 ISBN 90-247-1576-8
65. R.J. Brook: [*George*] *Berkeley's Philosophy of Science.* 1973 ISBN 90-247-1555-5
66. T.E. Jessop: *A Bibliography of George Berkeley.* With: *Inventory of Berkeley's Manuscript Remains* by A.A. Luce. 2nd revised and enlarged ed. 1973 ISBN 90-247-1577-6
67. E.I. Perry: *From Theology to History.* French Religious Controversy and the Revocation of the Edict of Nantes. 1973 ISBN 90-247-1578-4
68. P. Dibbon, H. Bots et E. Bots-Estourgie: *Inventaire de la correspondance (1631–1671) de Johannes Fredericus Gronovius* [1611–1671]. 1974 ISBN 90-247-1600-4
69. A.B. Collins: *The Secular is Sacred.* Platonism and Thomism in Marsilio Ficino's *Platonic Theology.* 1974 ISBN 90-247-1588-1

ARCHIVES INTERNATIONALES D'HISTOIRE DES IDÉES
*
INTERNATIONAL ARCHIVES OF THE HISTORY OF IDEAS

70. R. Simon (éd.): *Henry de Boulainviller.* Œuvres Philosophiques, Tome II. 1975 ISBN 90-247-1633-0
For Œuvres Philosophiques, Tome I *see under Volume 58.*
71. J.A.G. Tans et H. Schmitz du Moulin: *Pasquier Quesnel devant la Congrégation de l'Index.* Correspondance avec Francesco Barberini et mémoires sur la mise à l'Index de son édition des Œuvres de Saint Léon, publiés avec introduction et annotations. 1974 ISBN 90-247-1661-6
72. J.W. Carven: *Napoleon and the Lazarists (1804–1809).* 1974 ISBN 90-247-1667-5
73. G. Symcox: *The Crisis of French Sea Power (1688–1697).* From the *Guerre d'Escadre* to the *Guerre de Course.* 1974 ISBN 90-247-1645-4
74. R. MacGillivray: *Restoration Historians and the English Civil War.* 1974 ISBN 90-247-1678-0
75. A. Soman (ed.): *The Massacre of St. Bartholomew.* Reappraisals and Documents. 1974 ISBN 90-247-1652-7
76. R.E. Wanner: *Claude Fleury (1640–1723) as an Educational Historiographer and Thinker.* With an Introduction by W.W. Brickman. 1975 ISBN 90-247-1684-5
77. R.T. Carroll: *The Common-Sense Philosophy of Religion of Bishop Edward Stillingfleet (1635–1699).* 1975 ISBN 90-247-1647-0
78. J. Macary: *Masque et lumières au 18e [siècle].* André-François Deslandes, Citoyen et philosophe (1689–1757). 1975 ISBN 90-247-1698-5
79. S.M. Mason: *Montesquieu's Idea of Justice.* 1975 ISBN 90-247-1670-5
80. D.J.H. van Elden: *Esprits fins et esprits géométriques dans les portraits de Saint-Simon.* Contributions à l'étude du vocabulaire et du style. 1975 ISBN 90-247-1726-4
81. I. Primer (ed.): *Mandeville Studies.* New Explorations in the Art and Thought of Dr Bernard Mandeville (1670–1733). 1975 ISBN 90-247-1686-1
82. C.G. Noreña: *Studies in Spanish Renaissance Thought.* 1975 ISBN 90-247-1727-2
83. G. Wilson: *A Medievalist in the 18th Century.* Le Grand d'Aussy and the Fabliaux ou Contes. 1975 ISBN 90-247-1782-5
84. J.-R. Armogathe: *Theologia Cartesiana.* L'explication physique de l'Eucharistie chez Descartes et Dom Robert Desgabets. 1977 ISBN 90-247-1869-4
85. Bérault Stuart, Seigneur d'Aubigny: *Traité sur l'art de la guerre.* Introduction et édition par Élie de Comminges. 1976 ISBN 90-247-1871-6
86. S.L. Kaplan: *Bread, Politics and Political Economy in the Reign of Louis XV.* 2 vols., 1976 Set ISBN 90-247-1873-2
87. M. Lienhard (ed.): *The Origins and Characteristics of Anabaptism / Les débuts et les caractéristiques de l'Anabaptisme.* With an Extensive Bibliography / Avec une bibliographie détaillée. 1977 ISBN 90-247-1896-1
88. R. Descartes: *Règles utiles et claires pour la direction de l'esprit en la recherche de la vérité.* Traduction selon le lexique cartésien, et annotation conceptuelle par J.-L. Marion. Avec des notes mathématiques de P. Costabel. 1977 ISBN 90-247-1907-0
89. K. Hardesty: *The 'Supplément' to the 'Encyclopédie'.* [Diderot et d'Alembert]. 1977 ISBN 90-247-1965-8
90. H.B. White: *Antiquity Forgot.* Essays on Shakespeare, [Francis] Bacon, and Rembrandt. 1978 ISBN 90-247-1971-2
91. P.B.M. Blaas: *Continuity and Anachronism.* Parliamentary and Constitutional Development in Whig Historiography and in the Anti-Whig Reaction between 1890 and 1930. 1978 ISBN 90-247-2063-X

ARCHIVES INTERNATIONALES D'HISTOIRE DES IDÉES
*
INTERNATIONAL ARCHIVES OF THE HISTORY OF IDEAS

92. S.L. Kaplan (ed.): *La Bagarre.* Ferdinando Galiani's (1728–1787) 'Lost' Parody. With an Introduction by the Editor. 1979 ISBN 90-247-2125-3
93. E. McNiven Hine: *A Critical Study of [Étienne Bonnot de] Condillac's* [1714–1780] *'Traité des Systèmes'.* 1979 ISBN 90-247-2120-2
94. M.R.G. Spiller: *Concerning Natural Experimental Philosphy.* Meric Casaubon [1599–1671] and the Royal Society. 1980 ISBN 90-247-2414-7
95. F. Duchesneau: *La physiologie des Lumières.* Empirisme, modèles et théories. 1982 ISBN 90-247-2500-3
96. M. Heyd: *Between Orthodoxy and the Enlightenment.* Jean-Robert Chouet [1642–1731] and the Introduction of Cartesian Science in the Academy of Geneva. 1982 ISBN 90-247-2508-9
97. James O'Higgins: *Yves de Vallone* [1666/7–1705]: *The Making of an Esprit Fort.* 1982 ISBN 90-247-2520-8
98. M.L. Kuntz: *Guillaume Postel* [1510–1581]. Prophet of the Restitution of All Things. His Life and Thought. 1981 ISBN 90-247-2523-2
99. A. Rosenberg: *Nicolas Gueudeville and His Work (1652–172?).* 1982 ISBN 90-247-2533-X
100. S.L. Jaki: *Uneasy Genius: The Life and Work of Pierre Duhem* [1861-1916]. 1984 ISBN 90-247-2897-5; Pb (1987) 90-247-3532-7
101. Anne Conway [1631–1679]: *The Principles of the Most Ancient Modern Philosophy.* Edited and with an Introduction by P. Loptson. 1982 ISBN 90-247-2671-9
102. E.C. Patterson: *[Mrs.] Mary [Fairfax Greig] Sommerville* [1780–1872] *and the Cultivation of Science (1815–1840).* 1983 ISBN 90-247-2823-1
103. C.J. Berry: *Hume, Hegel and Human Nature.* 1982 ISBN 90-247-2682-4
104. C.J. Betts: *Early Deism in France.* From the so-called 'déistes' of Lyon (1564) to Voltaire's 'Lettres philosophiques' (1734). 1984 ISBN 90-247-2923-8
105. R. Gascoigne: *Religion, Rationality and Community.* Sacred and Secular in the Thought of Hegel and His Critics. 1985 ISBN 90-247-2992-0
106. S. Tweyman: *Scepticism and Belief in Hume's 'Dialogues Concerning Natural Religion'.* 1986 ISBN 90-247-3090-2
107. G. Cerny: *Theology, Politics and Letters at the Crossroads of European Civilization.* Jacques Basnage [1653–1723] and the Baylean Huguenot Refugees in the Dutch Republic. 1987 ISBN 90-247-3150-X
108. Spinoza's *Algebraic Calculation of the Rainbow & Calculation of Changes.* Edited and Translated from Dutch, with an Introduction, Explanatory Notes and an Appendix by M.J. Petry. 1985 ISBN 90-247-3149-6
109. R.G. McRae: *Philosophy and the Absolute.* The Modes of Hegel's Speculation. 1985 ISBN 90-247-3151-8
110. J.D. North and J.J. Roche (eds.): *The Light of Nature.* Essays in the History and Philosophy of Science presented to A.C. Crombie. 1985 ISBN 90-247-3165-8
111. C. Walton and P.J. Johnson (eds.): *[Thomas] Hobbes's 'Science of Natural Justice'.* 1987 ISBN 90-247-3226-3
112. B.W. Head: *Ideology and Social Science.* Destutt de Tracy and French Liberalism. 1985 ISBN 90-247-3228-X
113. A.Th. Peperzak: *Philosophy and Politics.* A Commentary on the Preface to Hegel's *Philosophy of Right.* 1987 ISBN Hb 90-247-3337-5; Pb ISBN 90-247-3338-3

ARCHIVES INTERNATIONALES D'HISTOIRE DES IDÉES
*
INTERNATIONAL ARCHIVES OF THE HISTORY OF IDEAS

114. S. Pines and Y. Yovel (eds.): *Maimonides* [1135-1204] *and Philosophy.* Papers Presented at the 6th Jerusalem Philosophical Encounter (May 1985). 1986 ISBN 90-247-3439-8
115. T.J. Saxby: *The Quest for the New Jerusalem, Jean de Labadie* [1610–1674] *and the Labadists (1610–1744).* 1987 ISBN 90-247-3485-1
116. C.E. Harline: *Pamphlets, Printing, and Political Culture in the Early Dutch Republic.* 1987 ISBN 90-247-3511-4
117. R.A. Watson and J.E. Force (eds.): *The Sceptical Mode in Modern Philosophy.* Essays in Honor of Richard H. Popkin. 1988 ISBN 90-247-3584-X
118. R.T. Bienvenu and M. Feingold (eds.): *In the Presence of the Past.* Essays in Honor of Frank Manuel. 1991 ISBN 0-7923-1008-X
119. J. van den Berg and E.G.E. van der Wall (eds.): *Jewish-Christian Relations in the 17th Century.* Studies and Documents. 1988 ISBN 90-247-3617-X
120. N. Waszek: *The Scottish Enlightenment and Hegel's Account of 'Civil Society'.* 1988 ISBN 90-247-3596-3
121. J. Walker (ed.): *Thought and Faith in the Philosophy of Hegel.* 1991 ISBN 0-7923-1234-1
122. Henry More [1614–1687]: *The Immortality of the Soul.* Edited with Introduction and Notes by A. Jacob. 1987 ISBN 90-247-3512-2
123. P.B. Scheurer and G. Debrock (eds.): *Newton's Scientific and Philosophical Legacy.* 1988 ISBN 90-247-3723-0
124. D.R. Kelley and R.H. Popkin (eds.): *The Shapes of Knowledge from the Renaissance to the Enlightenment.* 1991 ISBN 0-7923-1259-7
125. R.M. Golden (ed.): *The Huguenot Connection.* The Edict of Nantes, Its Revocation, and Early French Migration to South Carolina. 1988 ISBN 90-247-3645-5
126. S. Lindroth: *Les chemins du savoir en Suède.* De la fondation de l'Université d'Upsal à Jacob Berzelius. Études et Portraits. Traduit du suédois, présenté et annoté par J.-F. Battail. Avec une introduction sur Sten Lindroth par G. Eriksson. 1988 ISBN 90-247-3579-3
127. S. Hutton (ed.): *Henry More (1614–1687). Tercentenary Studies.* With a Biography and Bibliography by R. Crocker. 1989 ISBN 0-7923-0095-5
128. Y. Yovel (ed.): *Kant's Practical Philosophy Reconsidered.* Papers Presented at the 7th Jerusalem Philosophical Encounter (December 1986). 1989 ISBN 0-7923-0405-5
129. J.E. Force and R.H. Popkin: *Essays on the Context, Nature, and Influence of Isaac Newton's Theology.* 1990 ISBN 0-7923-0583-3
130. N. Capaldi and D.W. Livingston (eds.): *Liberty in Hume's 'History of England'.* 1990 ISBN 0-7923-0650-3
131. W. Brand: *Hume's Theory of Moral Judgment.* A Study in the Unity of *A Treatise of Human Nature.* 1992 ISBN 0-7923-1415-8
132. C.E. Harline (ed.): *The Rhyme and Reason of Politics in Early Modern Europe.* Collected Essays of Herbert H. Rowen. 1992 ISBN 0-7923-1527-8
133. N. Malebranche: *Treatise on Ethics* (1684). Translated and edited by C. Walton. 1993 ISBN 0-7923-1763-7
134. B.C. Southgate: *'Covetous of Truth'.* The Life and Work of Thomas White (1593–1676). 1993 ISBN 0-7923-1926-5
135. G. Santinello, C.W.T. Blackwell and Ph. Weller (eds.): *Models of the History of Philosophy.* Vol. 1: From its Origins in the Renaissance to the 'Historia Philosophica'. 1993 ISBN 0-7923-2200-2
136. M.J. Petry (ed.): *Hegel and Newtonianism.* 1993 ISBN 0-7923-2202-9

ARCHIVES INTERNATIONALES D'HISTOIRE DES IDÉES

*

INTERNATIONAL ARCHIVES OF THE HISTORY OF IDEAS

137. Otto von Guericke: *The New (so-called Magdeburg) Experiments* [Experimenta Nova, Amsterdam 1672]. Translated and edited by M.G. Foley Ames. 1994 ISBN 0-7923-2399-8
138. R.H. Popkin and G.M. Weiner (eds.): *Jewish Christians and Cristian Jews.* From the Renaissance to the Enlightenment. 1994 ISBN 0-7923-2452-8
139. J.E. Force and R.H. Popkin (eds.): *The Books of Nature and Scripture.* Recent Essays on Natural Philosophy, Theology, and Biblical Criticism in the Netherlands of Spinoza's Time and the British Isles of Newton's Time. 1994 ISBN 0-7923-2467-6
140. P. Rattansi and A. Clericuzio (eds.): *Alchemy and Chemistry in the 16th and 17th Centuries.* 1994 ISBN 0-7923-2573-7
141. S. Jayne: *Plato in Renaissance England.* 1995 ISBN 0-7923-3060-9
142. A.P. Coudert: *Leibniz and the Kabbalah.* 1995 ISBN 0-7923-3114-1
143. M.H. Hoffheimer: *Eduard Gans and the Hegelian Philosophy of Law.* 1995 ISBN 0-7923-3114-1
144. J.R.M. Neto: *The Christianization of Pyrrhonism.* Scepticism and Faith in Pascal, Kierkegaard, and Shestov. 1995 ISBN 0-7923-3381-0
145. R.H. Popkin (ed.): *Scepticism in the History of Philosophy.* A Pan-American Dialogue. 1996 ISBN 0-7923-3769-7
146. M. de Baar, M. Löwensteyn, M. Monteiro and A.A. Sneller (eds.): *Choosing the Better Part.* Anna Maria van Schurman (1607–1678). 1995 ISBN 0-7923-3799-9
147. M. Degenaar: *Molyneux's Problem.* Three Centuries of Discussion on the Perception of Forms. 1996 ISBN 0-7923-3934-7
148. S. Berti, F. Charles-Daubert and R.H. Popkin (eds.): *Heterodoxy, Spinozism, and Free Thought in Early-Eighteenth-Century Europe.* Studies on the *Traité des trois imposteurs.* 1996 ISBN 0-7923-4192-9
149. G.K. Browning (ed.): *Hegel's* Phenomenology of Spirit: *A Reappraisal.* 1997 ISBN 0-7923-4480-4
150. G.A.J. Rogers, J.M. Vienne and Y.C. Zarka (eds.): *The Cambridge Platonists in Philosophical Context.* Politics, Metaphysics and Religion. 1997 ISBN 0-7923-4530-4
151. R.L. Williams: *The Letters of Dominique Chaix, Botanist-Curé.* 1997 ISBN 0-7923-4615-7
152. R.H. Popkin, E. de Olaso and G. Tonelli (eds.): *Scepticism in the Enlightenment.* 1997 ISBN 0-7923-4643-2
153. L. de la Forge. Translated and edited by D.M. Clarke: *Treatise on the Human Mind (1664).* 1997 ISBN 0-7923-4778-1
154. S.P. Foster: *Melancholy Duty.* The Hume-Gibbon Attack on Christianity. 1997 ISBN 0-7923-4785-4
155. J. van der Zande and R.H. Popkin (eds.): *The Skeptical Tradition Around 1800.* Skepticism in Philosophy, Science, and Society. 1997 ISBN 0-7923-4846-X
156. P. Ferretti: *A Russian Advocate of Peace: Vasilii Malinovskii (1765–1814).* 1997 ISBN 0-7923-4846-6
157. M. Goldish: *Judaism in the Theology of Sir Isaac Newton.* 1998 ISBN 0-7923-4996-2
158. A.P. Coudert, R.H. Popkin and G.M. Weiner (eds.): *Leibniz, Mysticism and Religion.* 1998 ISBN 0-7923-5223-8
159. B. Fridén: *Rousseau's Economic Philosophy.* Beyond the Market of Innocents. 1998 ISBN 0-7923-5270-X
160. C.F. Fowler O.P.: *Descartes on the Human Soul.* Philosophy and the Demands of Christian Doctrine. 1999 ISBN 0-7923-5473-7

ARCHIVES INTERNATIONALES D'HISTOIRE DES IDÉES
*
INTERNATIONAL ARCHIVES OF THE HISTORY OF IDEAS

161. J.E. Force and R.H. Popkin (eds.): *Newton and Religion*. Context, Nature and Influence. 1999 ISBN 0-7923-5744-2
162. J.V. Andreae: *Christianapolis*. Introduced and translated by E.H. Thompson. 1999 ISBN 0-7923-5745-0
163. A.P. Coudert, S. Hutton, R.H. Popkin and G.M. Weiner (eds.): *Judaeo-Christian Intellectual Culture in the Seventeenth Century*. A Celebration of the Library of Narcissus Marsh (1638–1713). 1999 ISBN 0-7923-5789-2
164. T. Verbeek (ed.): *Johannes Clauberg* and Cartesian Philosophy in the Seventeenth Century. 1999 ISBN 0-7923-5831-7
165. A. Fix: *Fallen Angels*. Balthasar Bekker, Spirit Belief, and Confessionalism in the Seventeenth Century Dutch Republic. 1999 ISBN 0-7923-5876-7
166. S. Brown (ed.): *The Young Leibniz and his Philosophy (1646–76)*. 2000 ISBN 0-7923-5997-6
167. R. Ward: *The Life of Henry More*. Parts 1 and 2. 2000 ISBN 0-7923-6097-4
168. Z. Janowski: *Cartesian Theodicy*. Descartes' Quest for Certitude. 2000 ISBN 0-7923-6127-X
169. J.D. Popkin and R.H. Popkin (eds.): *The Abbé Grégoire and his World*. 2000 ISBN 0-7923-6247-0
170. C.G. Caffentzis: *Exciting the Industry of Mankind. George Berkeley's Philosophy of Money*. 2000 ISBN 0-7923-6297-7
171. A. Clericuzio: *Elements, Principles and Corpuscles*. A Study of Atomisms and Chemistry in the Seventeenth Century. 2001 ISBN 0-7923-6782-0
172. H. Hotson: *Paradise Postponed*. Johann Heinrich Alsted and the Birth of Calvinist Millenarianism. 2001 ISBN 0-7923-6787-1
173. M. Goldish and R.H. Popkin (eds.): *Millenarianism and Messianism in Early Modern European Culture*. Volume I. Jewish Messianism in the Early Modern World. 2001 ISBN 0-7923-6850-9
174. K.A. Kottman (ed.): *Millenarianism and Messianism in Early Modern European Culture*. Volume II. Catholic Millenarianism: From Savonarola to the Abbé Grégoire. 2001 ISBN 0-7923-6849-5
175. J.E. Force and R.H. Popkin (eds.): *Millenarianism and Messianism in Early Modern European Culture*. Volume III. The Millenarian Turn: Millenarian Contexts of Science, Politics and Everyday Anglo-American Life in the Seventeenth and Eighteenth Centuries. 2001 ISBN 0-7923-6848-7
176. J.C. Laursen and R.H. Popkin (eds.): *Millenarianism and Messianism in Early Modern European Culture*. Volume IV. Continental Millenarians: Protestants, Catholics, Heretics. 2001 ISBN 0-7923-6847-9
177. C. von Linné: *Nemesis Divina*. (edited and translated with explanatory notes by M.J. Petry). 2001 ISBN 0-7923-6820-7
178. M.A. Badía Cabrera: *Hume's Reflection on Religion*. 2001 ISBN 0-7923-7024-4
179. R.L. Williams: *Botanophilia in Eighteenth-Century France*. The Spirit of the Enlightenment. 2001 ISBN 0-7923-6886-X
180. R. Crocker (ed.): *Religion, Reason and Nature in Early Modern Europe*. 2001 ISBN 1-4020-0047-2

ARCHIVES INTERNATIONALES D'HISTOIRE DES IDÉES

*

INTERNATIONAL ARCHIVES OF THE HISTORY OF IDEAS

---

181. E. Slowik: *Cartesian Spacetime*. Descartes' Physics and the Relational Theory of Space and Motion. 2001 ISBN 1-4020-0265-3
182. R.L. Williams: *French Botany in the Enlightenment*..The Ill-fated Voyages of *La Pérouse* and His Rescuers. 2003 ISBN 1-4020-1109-1
183. A. Leshem: *Newton on Mathematics and Spiritual Purity*. 2003 ISBN 1-4020-1151-2

KLUWER ACADEMIC PUBLISHERS – DORDRECHT / BOSTON / LONDON